ISBN 978-1-5285-5101-4
PIBN 10030802

FB &c Ltd, Dalton House, 60 Windsor Avenue, London, SW19 2RR.
Company number 08720141. Registered in England and Wales.

For support please visit www.forgottenbooks.com

APPLIED CHEMISTRY;

IN

MANUFACTURES, ARTS, AND DOMESTIC ECONOMY.

EDITED BY

EDWARD ANDREW PARNELL,

AUTHOR OF "ELEMENTS OF CHEMICAL ANALYSIS," LATE ASSISTANT CHEMICAL LECTURER IN THE MEDICAL SCHOOL OF ST. THOMAS'S HOSPITAL, ETC.

PRELIMINARY OBSERVATIONS.
GAS ILLUMINATION.
PRESERVATION OF WOOD.
DYEING AND CALICO-PRINTING.

LONDON:
PRINTED FOR TAYLOR AND WALTON,
UPPER GOWER STREET.

1844.

London:
Printed by S. & J. Bentley, Wilson, and Fley,
Bangor House, Shoe Lane.

PREFACE.

The "Preliminary Observations," with which this volume commences, comprise some brief considerations on the fundamental doctrines of chemistry, most of which may be included in an account of the general laws which govern, and the phenomena and changes which accompany, the chemical combination of different bodies. This introduction concludes with an explanation of Chemical Symbols, and a table of Equivalents.

The article on Gas Illumination comprises not only a detailed account of the process of making light-gas from coal, and of the chemical and physical properties of the constituents of coal-gas, but several important collateral subjects; among which are some considerations on the modes of burning gas and on the economy of gas illumination, descriptions of the processes for making light-gas from other sources than coal, and an account of the applications which the secondary products of the coal-gas manufacture have received. The construction of different kinds of regulators and meters, and the process of "naphthalizing" gas, are also considered.

In the article on the Preservation of Wood, which follows that on Gas Illumination, means are pointed out for effectually guarding against the ordinary decomposing influences of air and water on timber. A considerable portion of the article is devoted to an account of the important experiments of Mr. W. Hyett, on the effects of the impregnation of wood with foreign substances.

Dyeing and Calico-printing forms the comprehensive subject of the concluding article of the present volume. Most of the leading processes now practised by the calico-printers of this country are there described, accompanied with explanations of the scientific principles on which they are based. Several of these processes, new to treatises on the subject, will be found of considerable value.

In many parts of the article on Calico-printing, I have derived material assistance from my esteemed friend Mr. Mercer, of the Oakenshaw print-works, near Blackburn, to whom I feel myself bound to return my sincere thanks. I am also desirous of expressing my obligations to Mr. John Graham of the Mayfield print-works, Manchester; and to Mr. James Hindle of the Sabden works, near Whalley.

TABLE OF CONTENTS.

	PAGE
PRELIMINARY OBSERVATIONS	1
Attainments of the ancients in some chemical arts.—Probably acquired by chance.—Processes mostly devised and improved now by deductions from principles.—Theoretical principles may be erroneous and the processes successful.—A few valuable processes discovered accidentally in modern times.—Amalgamation process of Hernando Velasquez	1
Chemical affinity	7
Elements.—Difference between a chemical compound and a mechanical compound	8
Chemical affinity exerted with different degrees of force between different substances	10
Chemical union of substances takes place in fixed and definite proportions.—The same element combines with different quantities of different substances.—Table of chemical equivalents of elements.—Bodies always unite chemically in the proportion of their equivalents	13
Heat developed through chemical combination . .	20
Chemical combination may be direct or indirect.—Double decomposition and double combination.—Single decomposition and single substitution	21
Solution.—Crystallization	24
Proximate and ultimate constituents	27
Symbols	30
Table of equivalents of acids and bases	34

I. GAS ILLUMINATION.

Outline of the process of making light-gas	35
§ I. HISTORICAL SKETCH OF THE PROGRESS OF GAS ILLUMINATION	37
§ II. COMPOSITION OF COAL	47
General composition.—Ultimate composition.—Atomic constitution of splint and cannel coal.—Extraneous matters . .	47

II. PRESERVATION OF WOOD.

PAGE

Rot and decay of wood are chemical changes.—Two conditions in which decay of wood can take place.—Nature of dry-rot . 151

§ I. PROPERTIES AND COMPOSITION OF WOOD . . 153

Woody tissue.—Cellular tissue.—Structure of the first year's stem of an exogen.—Manner of growth of exogens.—Medullary rays. —Duramen or heart-wood.—Alburnum or sap-wood.—Origin of of woody tissue.—Endogens 153

Seasoning of wood 159

Density of different kinds of wood 162

Soluble matters in wood.—Vegetable albumen.—Starch . . 164

Preparation and properties of pure lignin. — Dextrin. — Starch sugar.—Composition of oak-wood and beech-wood.—Researches of M. Payen on wood.—Cellulose.—Products of the destructive distillation of wood.—Flame of burning wood.—Wood smoke.—Heating power of different kinds of wood in a state of combustion.—Composition of wood ashes . . . 165

§ II. NATURE AND CAUSES OF THE DECAY OF WOOD 175

Decay of woody fibre a process of oxidation.—Eremecausis . 175

Decay of wood always produced by the contact of another body undergoing a similar change 181

Decay of wood closely analogous to acetification . . . 181

Albuminous matter in wood acts as a ferment . . . 182

Processes for ascertaining the proportion of albuminous matter in wood 183

§ III. PRESERVATIVE MATERIALS . . . 184

Decay of wood may be prevented by removing the albumen or by preventing its decomposition 184

Removal of the albumen by water 185

Corrosive sublimate 185

Action of corrosive sublimate confined to the albumen . . 186

Process for applying corrosive sublimate . . . 187

Protection of wood by the application of a varnish . . 187

List of preservative materials 188

Sulphate of copper ; sulphate of iron and sulphate of zinc . 189

Alkalies and alkaline earths 189

Alum 190

Common salt 190

Vegetable and animal oils 191

III. DYEING AND CALICO-PRINTING.

PAGE

Dyeing a chemical art.—Object of dyeing 221

§ I. History of Dyeing and Calico-printing . . . 222

Dyeing practised from time immemorial in the East.—Tyre celebrated for its dyers.—Tyrian purple.—Dyers of ancient Greece and Rome 222

Progress of dyeing in Europe affected by the invasion of the Northern barbarians in the 5th century.—The art did not revive until the 12th or 13th century.—Florentine dyers . . . 223

Progress of the art affected by the discovery of America.—Application of a salt of tin.—Use of logwood and indigo prohibited.—Madder introduced.—Application of mineral colours.—Turkey red 224

Proficiency of the ancients in topical dyeing.—Variegated linen cloths of Sidon.—Derivation of the term calico-printing.—Pliny's account of the Egyptian process of topical dyeing.—Pallampoors.—Topical dyeing in Mexico 226

Progress of calico-printing in Europe.—Printed cottons and linens of Augsburg.—First print-ground in England.—Calico-printing favoured by the prohibition of the importation of chintzes.—Duty imposed on printed calicos.—Wearing of all printed calicos prohibited in 1720.—A mixed fabric of linen and cotton was allowed to be printed in 1730, and the uniform cotton fabric in 1774.—Duty of threepence-halfpenny per square yard repealed in 1831 228

Mechanical improvements in calico-printing.—A copper-plate the first improvement on the wooden hand printing block.—Invention of cylinder or roller printing.—Capabilities of cylinder printing.—Surface printing.—Mule machine.—Press machine . . 230

Chemists who investigated the principles of dyeing during the last century.—Application of chlorine for bleaching.—Introduction of mineral colours.—Antimony orange and manganese bronze first introduced by Mr. Mercer, and the two chromates of lead by M. Kœchlin 232

§ II. General properties of Vegetable colouring matters 233

Only a few have been isolated in a pure state.—General composition.—Action of humid air.—Action of acids and alkalies.—Many organic colouring matters are feeble acids.—Lakes.—The attraction of colouring matters for insoluble bases is chemical affinity, and not a mere surface attraction 233

Action of chlorine, chromic acid, and sulphurous acid on organic colouring matters.—Animal charcoal.—Several colouring matters

APPLIED CHEMISTRY;

IN

MANUFACTURES, ARTS, AND DOMESTIC ECONOMY.

PRELIMINARY OBSERVATIONS.

REALLY valuable improvements of old and inventions of new processes in the chemical arts and manufactures are not often the result of chance, in the present advanced state of chemistry, but of a logical application of recognised chemical principles of greater or less generality. Such has been the case since the existence of chemistry as a distinct science possessed of general laws.

The surprising attainments of the ancients in certain chemical arts would lead us to infer that processes may be invented, and even successfully applied, without the smallest conception of the principles on which they depend. The ancient Egyptians and Phœnicians were not only acquainted with the means of extracting iron, copper, gold, silver, lead, and tin from the ores containing these metals, but with complicated processes for the preparation of several metallic compounds. They extracted *soda* from the soil in which that alkali naturally exists, and understood the means of purifying it: they procured *potash* from the ashes of plants, and made *soap* by combining the alkalies with oils and fats. They were acquainted with the mode of converting an alkaline carbonate into a caustic alkali by the action of quick-lime, and even took advantage of this property of lime to give to soda (carbonate of soda) a degree of causticity which deceived the purchasers of this article as to its real value. The arts of making earthenware, glass (both colourless and coloured), porcelain, and various pigments, and certain processes in dyeing, were brought to a state of perfection not exceeded, nay, in some instances, scarcely equalled, by artists of the present day.

These and other processes were not only invented without the aid of *science*, that term being accepted in its ordinary sense of inductive systematic truth, but they were also practised for ages before even an attempt was made to discover the causes of the varied phenomena presented in the different operations. The successive improvements fortunately achieved in some of the processes were contrived, like each original process itself, by chance, and consequently, in general, at the expense of great labour; the experimentalist being guided in his researches for improvements by no fixed principles or general laws. Where improvements are to be the result of accidental observations, they must of necessity be rarely made. The Chinese, it is true, practise certain chemical arts with surprising success: but these arts are no further advanced now than they have been for ages past; not because they have attained perfection, but from the want of a recognition of the principles on which the processes depend. In departing from, or even in merely modifying, the original process, there is a danger of losing it altogether. This is the reason, we apprehend, of the loss of some of the most refined processes of the ancients during the middle ages.

But since chemistry has existed as a science founded on certain laws, and since the properties of individual substances have been so minutely studied that the action of two or more bodies on each other in certain circumstances can be partly predicated before ascertained by experiment, new processes are devised and others improved by a scientific deduction from recognized principles, and from the known properties of the substances employed in the art or manufacture.

Refined as were some of the processes of the ancients, there are those practised in the present day which it is difficult to conceive could have been invented by any number of hazard experiments on substances presented by nature. They were devised and improved by a logical induction from known principles or admitted theories. The greater number of our calico-printing processes are illustrations of our assertion. For instance, we cannot imagine that any number of chance experiments would have led to the procuring from chrome-iron ore, (a black heavy mineral,) by the action, first of an alkali, and afterwards of an acid, of a substance by which, when applied in a proper manner, the calico-printer is enabled to produce a white pattern on a piece of cloth dyed with indigo-blue. This was, of course, discovered experimentally, but by experiments founded on scientific principles. Again, the common process for preparing carbonate of soda from common salt, invented towards the close of last century by Leblanc and others, which is scarcely second in importance to any chemical manufacturing process, could hardly have been devised otherwise than by a series of experiments conducted on certain chemical principles. This process consists, first, in the conversion of common salt to sulphate of soda by the action of sulphuric acid; in the next place, in the conversion of sulphate of soda to sulphuret of sodium by the action of carbonaceous matters at a high temperature; and, lastly, in the conversion of sulphuret of sodium to carbonate of soda by the action of chalk. The last two operations are now conducted simultaneously. Such a process as this we can hardly conceive to have been discovered by hazard experiments.

The theoretical principles on which a process is invented may be false, and yet the process itself turn out successful. Thus it was with the original process for obtaining sulphuric acid by the combustion of sulphur with nitre, which was contrived by Lefevre and Lemery soon after the developement, by Stahl and Beccher, of the *theory of phlogiston*. According to that theory all combustible bodies are compounds of *phlogiston*, or the principle of inflammability, (which is parted with in the act of combustion,) and some elementary basis, which is, in fact, the material product of the combustion. The latter is now known to be the compound body, being composed of the combustible matter and oxygen; the act of combustion is merely the combination of oxygen with the combustible. But consistently with the then prevailing theory of phlogiston, sulphur is a compound body, being composed of phlogiston united with (what is now termed) sulphuric acid: to procure the latter, therefore, it became necessary to withdraw phlogiston from the sulphur. Now nitre was considered to be a substance possessed of a dephlogisticating property in an eminent degree; and, when a mixture of sulphur and nitre was set on fire over water, sulphuric acid was found to be produced. Here the expected result was obtained, but the principle which led to the experiment was erroneous; for the nitre, instead of withdrawing anything from the sulphur, actually imparts oxygen to it: sulphuric acid being produced by the combination of the sulphur with a portion of the oxygen contained in the nitre. Such is a simple expression of the first changes in the process; afterwards they become somewhat more complicated. The reason why the process was successful, while the theory on which it was founded is false, is obvious. The terms a *dephlogisticating body* (on the theory of phlogiston), and an *oxidating body* (according to our present views), are synonymous: as are also a *dephlogisticated* and an *oxidated* substance.

It cannot be denied, however, that valuable processes and improvements have been devised of late years independently of chemical hypotheses and principles. The amalgamation process of Hernando Velasquez for obtaining silver from the ore (practised in Mexico) must ever be considered a wonderful instance of the discovery, by chance, of a process dependent in its details on complicated chemical agencies. The inventor was entirely unacquainted with theoretical chemistry, and was led to his process by experiment after experiment without connection, and guided, it would seem, by no general principles. But so refined is the process that its theory has only been developed of late years, with the united efforts of Gay-Lussac, Humboldt, Karsten, Boussingault, and Sonneschmidt. Instances like this are, however, very rare.

Although many processes in the chemical arts are dependent for their successful practice chiefly on a knowledge of principles, yet, in general, the ambition of those engaged in working the process is to attain dexterity in the mechanical operations, without the smallest endeavour to comprehend the most obvious principles of their process. Although chemistry is making rapid advances, many of the arts dependent on the science are stationary, the artisan or manufacturer being, generally, too ignorant of chemistry to know where to look for improvements. By this neglect of the practical man, scientific chemistry is also much retarded. In a dye-house and a calico-print-work, for instance, how many interesting phenomena pass unobserved, or at least without being investigated, which, if carefully examined by a scientific chemist, might prove the sources of brilliant discoveries, conducive no less to the advancement of theoretical chemistry than of the chemical art itself!

To consider minutely the general principles of chemical philosophy would be inconsistent with the plan of the present work. But for facility of reference, and for the convenience of those who have not especially studied the principles of theoretical chemistry, it has been considered proper to commence with a few observations on the nature of chemical affinity, and on the laws which govern, and the phenomena which accompany, the chemical combination of different substances. All other important theoretical considerations affecting the various practical subjects treated of in this work will be adverted to under the respective articles.

CHEMICAL AFFINITY.

The vast variety of substances by which we are surrounded, animate and inanimate, solid, liquid, and gaseous, are composed of, comparatively, a very small number of *elements* or *simple substances;* of bodies, that is, which contain one kind of matter only. According to the present state of our knowledge, the number of elements does not exceed fifty-six; and, as the only evidence we can obtain of the elementary character of a substance is of a negative kind, namely, that we are unable now to decompose it, or resolve it into other kinds of matter, it is not unlikely that some of the substances now considered as elements may hereafter turn out to be compound. But until such is proved to be the case, it is proper to consider them as simple substances.

Every substance in nature, then, is either an element, or else it results from the combination of two or more elements with each other. This union of the elements does not take place indiscriminately: some of them when brought into contact exhibit a remarkable proneness to unite, while others may be mixed most intimately without the occurrence of combination. Peculiar attachments and indifferences certainly subsist between different substances. The former may be conveniently distinguished as chemical affinity, or chemical attraction. The exertion of this force is not confined to simple substances merely; compound bodies, as sulphuric acid and water, for instance, are equally subject to its influence.

When the constituents of a compound body are thus held together by chemical affinity, the resulting substance is called a *chemical compound*, to distinguish it from a *mechanical compound.* The latter is a mere mixture of different substances, which are held together by no chemical affinity, but, if the body is a solid or a liquid, merely by the force of cohesion or of adhesion. Chemical compounds are definite combinations of two or more elements, united in a particular manner, and always, when forming the same compound, in fixed proportions; the substance formed always wants the chemical characters of its constituents, and in the act of combination there is commonly a very obvious change, even in the most general, of physical properties. Mechanical compounds, which constitute by far the greater number of natural substances, are on the contrary indefinite mixtures of elementary bodies or chemical compounds of the elements, retaining both the form and the properties of their constituents. The atmosphere around us, for example, is a mechanical mixture of two elementary gases, oxygen and nitrogen. These same elements when chemically united in a certain proportion give rise to one of the most corrosive of all acids, namely, nitric acid, or aqua-fortis as it is popularly designated.

Again, the rock granite is a mechanical mixture of three chemical compounds: 1°, *quartz*, which is a chemical combination of two elements, silicon and oxygen; 2°, *felspar*, which is composed of four elements, silicon, oxygen, potassium, and aluminum; and, 3°, *mica*, composed of six elements, namely, silicon, oxygen, iron, aluminum, magnesium, and potassium. The three minerals, quartz, felspar, and mica, each a chemical combination, thus constitute the rock granite by their mechanical admixture.

The nature of chemical affinity and the characters of chemical compounds may be conveniently considered under the four following heads: — 1°, Chemical affinity is exerted with different degrees of force between different substances. 2°, The union takes place in certain fixed and definite proportions. 3°, The resulting compound always differs essentially in chemical, and often in physical properties from its constituents; and 4°, The union may take place indirectly by substitution, or directly.

1. *Chemical affinity is exerted with different degrees of force between different substances.*—On examining various metallic oxides it is found that the oxygen and metal are united together with very different degrees of intensity in the different oxides. Thus, oxide of silver is easily decomposed when heated to redness, oxygen gas being given off and metallic silver remaining behind. Oxide of copper, on the contrary, is not reduced to the metallic state at the highest temperature to which we can expose it. It is therefore reasonably inferred that oxygen has a more powerful affinity for copper than for silver. Again, oxide of copper is reduced to the state of metallic copper with the greatest facility if heated to redness in a glass tube through which a stream of hydrogen gas is made to pass. But potash, the oxide of potassium, does not undergo the smallest decomposition under such circumstances. Hence oxygen has a more powerful affinity for potassium than for copper.

The truth of the proposition is clearly exhibited in the decomposition of a compound substance by another body. On introducing metallic copper into a solution of chloride of mercury (corrosive sublimate), the latter is decomposed, metallic mercury is precipitated, and a proportional quantity of copper is dissolved, uniting with chlorine to form chloride of copper. But chloride of copper may be decomposed in a similar manner by metallic tin, and chloride of tin by metallic zinc. The relative affinity of these metals for chlorine may therefore be arranged in the following order,—

Chlorine.

1. Zinc,	3. Copper,
2. Tin,	4. Mercury:

a solution of chloride of zinc is not decomposed by any of the metals mentioned, while a solution of chloride of mercury is decomposed by each.

A series of tables was contrived by Geoffroy and Bergmann, founded on such experiments as these, intended to exhibit the order of affinity of different *bases* * for an *acid*, and of different acids for a base. The name of an acid or a base is put at the head of a list, and after it the names of all the bases or acids in the order of their affinity. For example, the affinity of alkalies and earths for sulphuric acid is expressed as follows:—

Sulphuric acid.

1. Barytes,	5. Lime,
2. Strontian,	6. Magnesia,
3. Potash,	7. Ammonia.
4. Soda,	

Barytes, the first in the list, is capable of taking sulphuric acid from the sulphate of either of the bases which follow, while sulphate of barytes is not decomposed by any other base. Potash has a stronger affinity for sulphuric acid than the bases which follow, but not so strong as that of barytes. The experiments are supposed to be made on aqueous solutions of the several sulphates.

* By a base is meant an alkali, or a metallic oxide which has a tendency to unite with an acid and thus form a *salt*. Thus potash is the base of nitre, or nitrate of potash; oxide of lead is the base of sugar of lead, or acetate of lead.

Such tables, however, shew the order of decomposition, and the relative force of affinity, in one set of conditions only; for, when the same two substances are brought together under different circumstances, very different reactions may ensue. Of this Bergmann himself was aware, and hence gave two lists or columns in some of his tables of affinity; one shewing the order of decomposition when the substances are mixed together dissolved in water, as above, and the other when mixed in the dry state and strongly heated. Boracic acid possesses very feeble acid characters when dissolved in water; it is hardly able to destroy the alkaline properties of potash or soda; but at a red heat, from its great fixedness, it comports itself as a most powerful acid. If dry nitrate of ammonia and carbonate of lime are mixed and exposed to heat, a decomposition takes place with production of nitrate of lime and carbonate of ammonia; from which circumstance it would be inferred that nitric acid has a stronger affinity for lime than for ammonia, or carbonic acid a stronger affinity for ammonia than for lime. But if the conditions in which the nitric acid, carbonic acid, lime, and ammonia are brought together are altered, very different results take place. On mixing aqueous solutions of nitrate of lime and carbonate of ammonia, carbonate of lime is precipitated, and nitrate of ammonia remains in solution. Again, carbonate of potash dissolved in *water* is instantly decomposed on the addition of acetic acid, carbonic acid gas being evolved and acetate of potash formed. Acetic acid manifests in this case a stronger affinity for potash than carbonic acid does. But if a stream of carbonic acid gas is passed through a solution of acetate of potash dissolved in *alcohol*, the carbonic acid seems to be the most powerful, for carbonate of potash is formed and acetic acid liberated. The result in the latter case seems to depend on the insolubility of carbonate of potash in alcohol. If the composition of two salts is such as to permit the formation of an insoluble product by their mutual decomposition, (as in the case of carbonate of ammonia and nitrate of lime,) such a change is sure to take place when solutions of the salts are mixed. The order of decomposition seems to be decided, in such a case, not by the superior affinity of this or that acid or base, but by the tendency to the formation of an insoluble substance.

These and many other examples, which might be adduced if necessary, shew that though chemical affinity is exerted between different bodies with different degrees of force, yet accessory circumstances do sometimes materially affect the relative affinities of different substances.

2. *The chemical union of substances takes place in certain fixed and definite proportions.*— The circumstance, that the proportions in which two or more bodies can unite chemically are limited on both sides, has been already alluded to as a distinction between a chemical and a mechanical compound (page 8). Common salt, for instance, by whatever process it may be formed, is iuvariably composed of 60 parts of sodium, and 40 parts of chlorine; and, so far as is known, these elements are incapable of uniting in any other proportion. Pure

carbonate of lime, whether formed ages ago by the hand of nature, or quite recently by the chemist, whether in the form of white marble, chalk, or a stalactite, whether in the six-sided prism of arragonite, or in the rhomboidal crystal of calc spar, is always composed of 43·7 parts of carbonic acid and 56·3 parts of lime. That the composition of chemical combinations is always fixed and invariable, is constantly borne out by analyses; and from this definite nature of chemical combination is derived in a great measure the character of chemistry as an exact science.

The same element combines with very different quantities of different substances. This is shewn in the following table of the composition of water, sulphuretted hydrogen, hydrochloric acid, and hydriodic acid, each of which substances contains hydrogen as a common constituent.

Water.		*Sulphuretted Hydrogen.*	
Hydrogen	11·1	Hydrogen	5·9
Oxygen	88·9	Sulphur	94·1
	100·0		100·0

Hydrochloric Acid.		*Hydriodic Acid.*	
Hydrogen	2·7	Hydrogen	0·79
Chlorine	97·3	Iodine	99·21
	100·0		100·00

By such a mode of expressing the composition of these bodies no simple relation is perceptible between the respective quantities of hydrogen in the different compounds. But if a constant number, say 1, is given to hydrogen, then it will be seen that 8 parts of oxygen, 16 parts of sulphur, 35·5 parts of chlorine, and 126 parts of iodine combine respectively with 1 part of hydrogen. Now the numbers mentioned are in some degree characteristic of the substances to which they are attached; for, on examining the composition of the compounds which these elements form with lead, copper, and potassium, it is found that the above numbers of parts of oxygen, chlorine, &c. actually combine with the same quantity of the metal, although the amounts of the different metals in the respective compounds are not the same. Thus

8 parts of oxygen		1 part of hydrogen,
35·5 parts of chlorine	combine with	104 parts of lead,
16 parts of sulphur		32 parts of copper,
126 parts of iodine		39 parts of potassium.

By extending this inquiry to other substances, a series of numbers may be obtained which exhibits either the exact relative quantities in which bodies unite to form chemical combinations, or else multiples of such quantities. These numbers are called *combining proportions* or *equivalents*. The latter term is employed to signify that the combining proportion of one body has the same value, or is, in one sense, equivalent to that of another body, and may be substituted for it in combination. The same quantity of potassium which unites with 35·5 parts of chlorine also combines with 126 parts of iodine; therefore the numbers 35·5 for chlorine, and 126 for iodine, are as it were equivalent to each other, as they have an equal power in combining with, and altering the nature of, 39 parts of potassium.

Hitherto we have considered the equivalent numbers with reference to hydrogen as unity, but this is quite arbitrary, and any other series of numbers may be used, provided the proper relation between them is preserved. Thus it is the practice of some chemists to form the series with reference to oxygen as 100, and others make oxygen 1. The series of numbers in which hydrogen is considered as 1 is called the hydrogen scale of equivalents, the others are called the oxygen scales. The hydrogen scale will be employed in the present work. The following table contains a list of all the elements which have been discovered up to the present time, with their combining numbers on both the hydrogen and oxygen scales.

TABLE OF CHEMICAL EQUIVALENTS.

Names of Elements.	Equivalents.		Names of Elements.	Equivalents.	
	Oxygen = 100.	Hydrogen = 1.		Oxygen = 100.	Hydrogen = 1.
Aluminum	171·2	13·7	Columbium	2307·4	184·9
Antimony	1612·9	129·0	Copper	395·7	31·7
Arsenic	940·1	75·3	Didymium	?	?
Barium	856·9	68·7	Fluorine	233·8	18·7
Bismuth	886·9	71·1	Glucinum	331·3	26·5
Boron	136·2	10·9	Gold	2486·0	199·2
Bromine	978·3	78·4	Hydrogen	12·5	1·0
Cadmium	696·8	55·8	Iodine	1579·5	126·6
Calcium	256·0	20·5	Iridium	1233·5	98·8
Carbon	75·0	6·0	Iron	339·2	27·2
Cerium	574·7	46·05	Lantanum	?	?
Chlorine	442·6	35·5	Lead	1294·5	103·7
Chromium	351·8	28·2	Lithium	80·3	6·4
Cobalt	369·0	29·6	Magnesium	158·3	12·7

Names of Elements.	Equivalents. Oxygen = 100.	Equivalents. Hydrogen = 1.	Names of Elements.	Equivalents. Oxygen = 100.	Equivalents. Hydrogen = 1.
Manganese	345·9	27·7	Silver	1351·6	108·3
Mercury	1265·8	101·4	Sodium	290·9	23·3
Molybdenum	598·5	48·0	Strontium	547·3	43·8
Nickel	369·7	29·6	Sulphur	201·17	16·1
Nitrogen	175·0	14·0	Tellurium	801·76	64·2
Osmium	1244·5	99·7	Thorium	744·9	59·8
Oxygen	100·0	8·0	Tin	735·29	58·9
Palladium	665·9	53·4	Titanium	303·66	24·3
Phosphorus	392·3	31·4	Tungsten	1183·0	94·8
Platinum	1233·5	98·8	Vanadium	856·9	68·7
Potassium	489·9	39·2	Uranium	2711·4	217·3
Rhodium	651·4	52·2	Yttrium	402·5	32·2
Selenium	494·6	39·6	Zinc	403·2	32·3
Silicon	277·3	22·2	Zirconium	420·2	33·7

The principal law which governs chemical combinations is, that bodies unite chemically with each other *only in the proportion of their equivalents*, or in *multiples of their equivalents*, and in no intermediate proportions. This law is well illustrated by the five compounds which are formed by the union of oxygen and nitrogen. The first of these, *protoxide of nitrogen*, or *nitrous oxide*, contains 14 parts of nitrogen and 8 of oxygen; that is, one equivalent of each constituent. The second combination of these elements, *deutoxide of nitrogen*, or *nitric oxide*, contains 14 parts of nitrogen and 16 of oxygen; that is, two equivalents of oxygen and one of nitrogen. The third, *nitrous acid*, contains 14 parts of nitrogen and 24 of oxygen, or three equivalents of oxygen and one of nitrogen. The fourth combination, *peroxide of nitrogen*, contains 14 parts of nitrogen or one equivalent, and 32 parts of oxygen or four equivalents. The fifth and last, *nitric acid*, of which *aqua-fortis* of commerce is a solution in water, is composed of 14 parts of nitrogen or one equivalent, and 40 parts of oxygen or five equivalents. In the five compounds, therefore, one eq. of nitrogen is combined respectively with one, two, three, four, and five equivalents of oxygen.

The law illustrated in the last paragraph applies equally to combinations among bodies themselves compound.

The equivalent of a compound body is always the sum of the equivalents of its constituents. Thus the equivalent of water on the hydrogen scale is 9; water being composed of one eq. of hydrogen = 1, and one eq. of oxygen = 8. The equivalent of potash is 47; that alkali containing one eq. of potassium = 39, and one eq. of oxygen = 8. Now 47 parts of potash combine with exactly 9 parts of water to form the hydrate of potash or compound of potash and water. As sulphuric acid is composed of one equivalent of sulphur = 16, and three equivalents of oxygen = 24, the equivalent of sulphuric acid is 40. To form neutral sulphate of potash, consisting of one eq. of potash and one eq. of sulphuric acid, 47 parts of potash combine with exactly 40 parts of sulphuric acid, and the equivalent of the resulting sulphate of potash is 87. An equivalent of sulphuric acid unites in like manner with 9, 18, and 27 parts of water to form hydrates containing respectively one, two, and three equivalents of water. Here is, therefore, an example of combination in multiple proportions between bodies themselves compound. However great may be the number of the equivalents of the elements of a compound body, the universality of the law that the equivalent of the compound is the sum of the equivalents of its constituents is by no means affected.

3. *The resulting compound always differs in chemical, and often in physical properties, from its constituents.*—A change in the properties of the combining substances is the leading circumstance that distinguishes chemical combination from mechanical admixture. In the former, the change is never wanting; in the latter, the properties of the mixture are the mean of those of its constituents. We should not suppose from its appearance that common salt is a compound body: much less that this harmless substance is composed, on the one part, of a body which when uncombined is a highly irritating and noxious gas (chlorine), which causes instant death if inhaled; and, on the other part, of a metal (sodium) as bright as silver, which has the extraordinary property of taking fire when moistened with water.

Sometimes a change in density and form attends the chemical combination of two substances. Thus two gases may give rise to a solid, as happens when ammoniacal gas and muriatic acid gas are mixed together. A solid sometimes assumes the gaseous state by combining with a gas; thus sulphur and charcoal form gases by combining with hydrogen gas. Gases sometimes unite, however, without any alteration in bulk. When combination takes place among solids and liquids with the formation of a solid or liquid, condensation is a more frequent result than expansion; and, if other evidence of *chemical* combination is wanting, that of the occurrence of a condensation may be considered conclusive. Thus the contraction which results when water and alcohol are mixed in certain proportions, is considered, of itself, a sufficient proof of the chemical union of the water with the alcohol.

A change of colour very frequently attends chemical combination ; sulphur forms a red and a black compound with mercury, black compounds with lead and copper, a white compound with zinc, and two yellow compounds with arsenic. Iodine, whose vapour is violet-coloured, forms a white compound with copper, a yellow compound with lead, and a scarlet compound with mercury. An example of a change in properties is afforded, in short, by almost every case of chemical combination.

Change of temperature is a frequent result of chemical action. It is said that the direct chemical combination of two or more substances never occurs without the developement of a certain amount of heat, on which circumstance depend all our ordinary artificial modes of obtaining heat. The production of heat in the act of combination may be sometimes referred partly to the condensation which usually takes place, the compound formed having a smaller specific heat than its constituents; but in a great many cases this explanation is inadmissible. The true cause of the evolution of heat in chemical combination still remains to be accounted for. The *intensity* of the heat developed through the combination of two bodies differs greatly, according to the circumstances in which the combination takes place; but it is a remarkable fact that the *quantity* or *absolute amount* of heat (to apply to an imponderable agent terms which belong to material substances) is the same for like quantities of the combining substances, whether the union takes place rapidly or slowly. Oxygen and hydrogen gases may be made to unite and form water quickly, as by contact with flame or spongy platinum; and slowly, by placing the mixed gases in contact with some earthy mould. In the first case, the intensity of the heat produced is sufficient to fuse with facility many substances which are quite infusible at the highest temperature of a smith's forge: in the second case, the temperature rises very little above that of the surrounding atmosphere. But there is reason to believe that the actual quantity of heat developed is the same in both cases for like quantities of the gases: the difference is only in intensity or state of concentration; the slowness of the action in the second case allowing the distribution of the heat to surrounding bodies to a far greater extent than could possibly be the case in the rapid combustion of the gases. The superior vivacity of the combustion of tallow, a match of wood, &c. in oxygen gas, compared with air, depends entirely on the same circumstance. A piece of charcoal burns with far more heat and light in oxygen gas than in air; but it unites with oxygen, or is consumed, proportionally faster.

4. *Chemical union may take place indirectly by substitution, or directly.*—The following examples will serve to illustrate this proposition. When

the two component parts of common salt, chlorine and sodium, are placed in contact, a direct combination instantly ensues. But the union of the chlorine and sodium can be effected in a very different manner, with the formation of the same compound. The combination takes place indirectly, for instance, when hydrochloric or muriatic acid, (a compound of hydrogen and chlorine,) is brought into contact with soda (a compound of sodium and oxygen). Again, if a mixture of iron filings and flowers of sulphur is projected into a crucible heated to dull redness, a direct combination of iron and sulphur ensues, the sulphuret of iron being formed. But the same compound is formed when a solution of sulphuret of potassium (a compound of sulphur and potassium) is mixed with a solution of chloride of iron (a compound of iron and chlorine).

Indirect combinations of this nature are characterized by two important circumstances. While union is taking place on the one hand, there is necessarily decomposition on the other; for the substances which are to unite are brought into contact in a state of previous combination with other substances. In the formation of common salt or chloride of sodium by hydrochloric acid and soda, there must be a decomposition both of the hydrochloric acid and the soda, in order that the chlorine and sodium may unite. And when sulphuret of iron is formed by mixing sulphuret of potassium with chloride of iron, the sulphur and iron cannot unite without a decomposition of the sulphuret of potassium and of the chloride of iron. As there are two disunions, so are there likewise, generally, two combinations. When hydrochloric acid is made to act on soda, union is effected, not only between the chlorine and sodium, but the hydrogen of the hydrochloric acid unites at the same moment with the oxygen of the soda, to produce water. Chloride of sodium and water, therefore, are the products of the action of hydrochloric acid on soda. The manner in which the decompositions and combinations occur is more clearly exhibited in the subjoined diagram. The first column contains the names of the original substances; the second, the names of their constituents; and the third, the names of the compounds which are formed, the lines indicating how the constituents of the original substances become arranged.

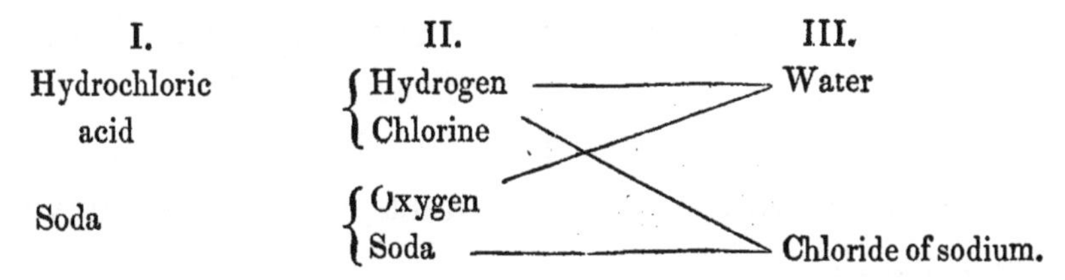

In like manner, as two decompositions occur when sulphuret of potassium and chloride of iron are made to act on each other, so are there likewise two combinations; for at the same time that the sulphur of the sulphuret of potassium unites with the iron of the chloride of iron, the chlorine of the latter unites with the potassium of the sulphuret of potassium to form chloride of potassium. The annexed diagram will illustrate these changes.

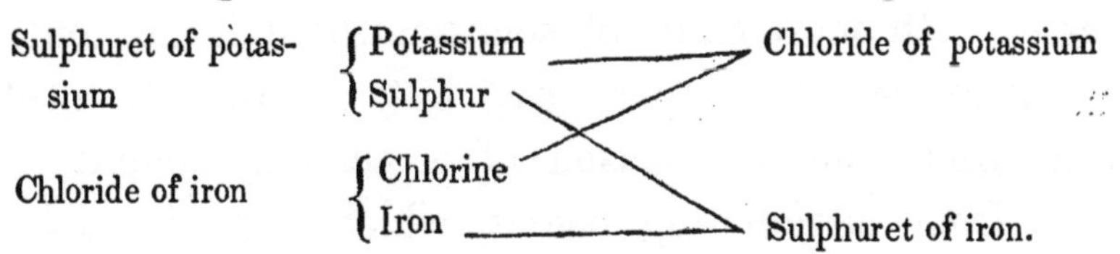

Such reactions as these are described as double decompositions or double substitutions. But another kind of indirect union frequently occurs, in which there is a single decomposition and a single substitution, only one of the two substances which are to be united being then in a state of previous combination with another substance. The action of iodine on sulphuretted hydrogen, explained in the annexed diagram, is of this nature.

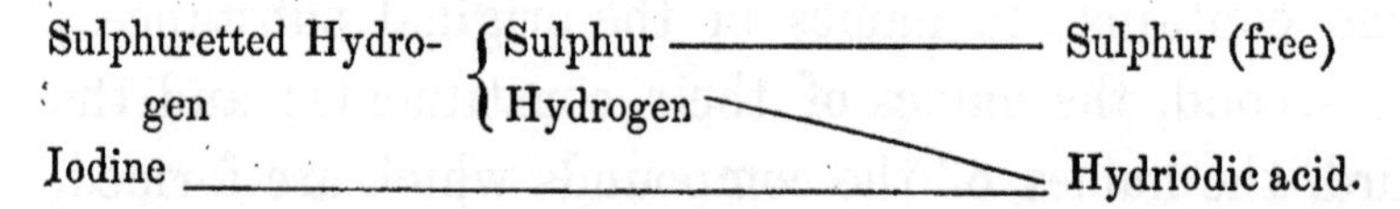

In this example there is a single union and a single decomposition; or, in fact, merely a substitution of iodine for sulphur in the sulphuretted hydrogen. A great number of compounds may be readily formed by reactions of this kind, all attempts to obtain which by direct combination have been fruitless.

Solution.— The attractive force by which one body is dissolved in another in the liquid state has been considered by some to be identical with chemical affinity. One of the characters of a chemical union is certainly in some degree fulfilled in solution, namely, change in properties; but several circumstances manifest a difference between the two forces. In most cases of solution there is not only no increase of temperature, but a production of cold from the abstraction of a certain amount of heat from surrounding bodies, which is rendered *la-*

tent by the dissolved substance, as always happens when a body passes from the solid to the liquid condition. In solution, the union takes place in no fixed proportions: a certain quantity of water can dissolve or combine with any quantity of common salt less than 37 per cent. of the weight of the water; and the solution has the same appearance, whether it contains a large or a small proportion of salt. Water, it is true, has a constant maximum solvent power for every soluble salt at a particular temperature; but the quantity of water and the quantity of salt in solution bear, in general, no simple relation to the chemical equivalents of these substances. Again, chemical combination takes place with so much the more energy as the properties of the combining substances are more opposed; but solution, on the contrary, with a readiness corresponding to their similarity in properties. Thus, to dissolve combustible bodies containing a large proportion of carbon and hydrogen, such as resin, fixed oils, essences, and caoutchouc, combustible liquids must be used, as alcohol, ether, naphtha, and oil of turpentine. To dissolve a metal, another metal must be used, as mercury:* oxidated bodies, as most salts, require oxidated solvents, as water and acids: very few saline matters soluble in water are also soluble in ether and in oils. An essential difference exists, therefore, between the act of solution and that of chemical combination; the former be-

* When a metal is said to be dissolved in an acid, as copper in nitric acid, for example, it is, in reality, not the metal itself which dissolves, but a substance formed by the chemical action of the acid on the metal, as nitrate of copper.

ing exerted by preference between similar, and the latter between dissimilar particles.

With a single exception, the solubility of a salt varies with the temperature of the solvent, the largest amount being taken up at the higher temperature. The exception to this law referred to is in the case of common salt, of which the same proportion is dissolved by water at all temperatures below 230° Fahrenheit. In another case, that of sulphate of soda, the solubility of the salt increases rapidly up to the temperature 92° Fahrenheit, at which it is at the maximum: the solution at that temperature contains 52 parts of dry sulphate to 100 parts of water, but at a higher temperature water is incapable of dissolving so large a proportion of that salt. In all other cases, so far as is known, the amount of a salt dissolved increases with the temperature, unless a decomposition and formation of a less soluble substance take place under the influence of the heat.

Advantage is taken of the increased solubility of salts at a high temperature to effect their crystallization. When a solution of a salt *saturated* (containing its maximum quantity of the salt) at a high temperature is allowed to cool, it retains in solution only that quantity proper to its reduced temperature: the excess is deposited, it may be in the state of crystals. Thus a saturated solution of Epsom salts at the temperature of 212° contains 74 parts of that salt to every 100 parts of water; but, on being cooled down to 50° Fahrenheit, 44 parts are deposited in crystals, 100 parts of water at 50° being incapable of holding in solution more than 30 parts of Epsom salts.

Some cases of solution, however, are undoubtedly accompanied with the chemical union of the solvent with the body dissolved ; as, for example, when dry potash or dry chloride of calcium is dissolved in water. In these, as in other cases of direct chemical combination, there is developement of heat, while cold is produced in a case of solution merely. When such a combination takes place, it is, strictly speaking, the compound formed which dissolves; as, the hydrate of potash, or the hydrate of chloride of calcium.

Proximate and ultimate constituents, and Symbols.—The composition of a compound body may be expressed in the simple elements which form what are called its *ultimate constituents*, and also very frequently by those compound bodies which by their immediate union compose the substance in question. The latter are called *proximate constituents.* Of dry sulphate of magnesia, for instance, the ultimate constituents are sulphur, oxygen, magnesium: the proximate constituents are sulphuric acid and magnesia: thus,

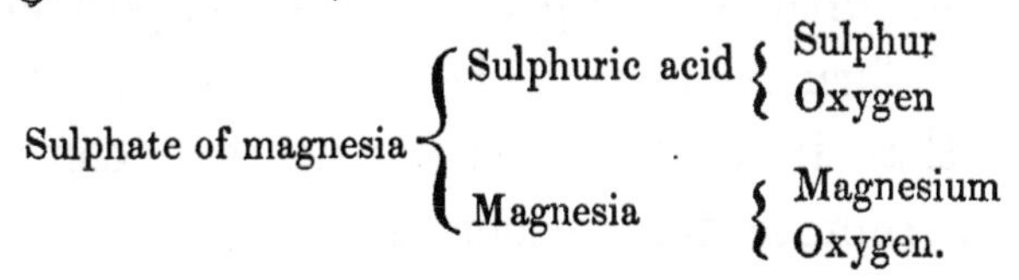

The composition of crystallized alum is more complex: thus,

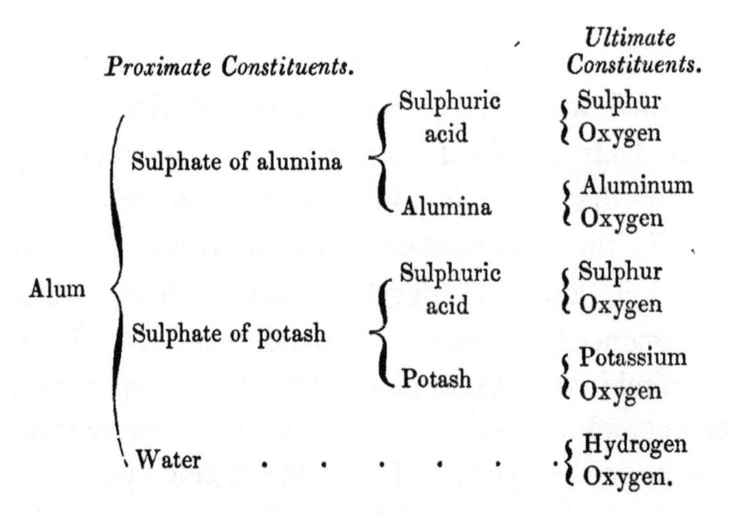

Now the composition of crystallized alum may be expressed in so many parts of sulphur, oxygen, aluminum, potassium, and hydrogen; or in so many parts of sulphuric acid, potash, alumina, and water; or in so many parts of sulphate of alumina, sulphate of potash, and water.

The expression of the composition of a substance which is a chemical, and not a mere mechanical compound, should include, if possible, the *atomic composition* of the substance, that is, the relative proportions of the equivalents of its constituents, as well as the ultimate or per-centage composition. Much more information may be conveyed in such a statement than in that of the per-centage compositiou alone, represented either according to the proximate or ultimate constituents. The reasons of this will be obvious from the considerations offered at page 13 *et seq.*, on the definite proportions in which chemical union takes place. As an example, the composition of crystallized sulphate of magnesia may be represented thus:

	Equivalent or atom.	*Atomic weight.*	*Per-centage composition.*
Magnesia	1	21	16·1
Sulphuric acid	1	40	32·1
Water	7	63	50·8
Sulphate of magnesia	1	124	100·0

The readiest method of representing the composition of a compound body is by the association of symbols of its elementary or proximate constituents. Each elementary substance is represented by a particular symbol, which is the initial letter of its Latin name; but when the names of two or more elements begin with the same letter, the distinction is made by an additional letter in a smaller character. Thus, phosphorus being represented by the letter P, platinum is indicated by Pt, and palladium by Pd; C being the symbol for carbon, calcium is represented by Ca, chlorine by Cl, and cobalt by Co; the small letter is significant only when in conjunction with the large letter. The following table includes the symbols of all the elementary substances known.

TABLE OF SYMBOLS.

Elements.	Symb.	Elements.	Symb.
Aluminum	Al	Manganese...............	Mn
Antimony (Stibium) ...	Sb	Mercury (Hydrargyrum)	Hg
Arsenic	As	Molybdenum	Mo
Barium	Ba	Nickel	Nk
Bismuth..................	Bi	Nitrogen	N
Boron......................	Bo	Osmium	Os
Bromine............. ...	B	Oxygen	O
Cadmium	Cd	Palladium	Pd
Calcium	Ca	Phosphorus	P
Carbon	C	Platinum	Pt
Cerium	Ce	Potassium (Kalium) ...	K
Chlorine.................	Cl	Rhodium	R
Chromium	Cr	Selenium	Se
Cobalt	Co	Silicon	Si
Columbium (Tantalum)	Ta	Silver (Argentum)	Ag
Copper (Cuprum)	Cu	Sodium (Natrium)......	Na
Didymium	Di	Strontium	Sr
Fluorine.................	Fl	Sulphur	S
Glucinum	Gl	Tellurium	Te
Gold (Aurum)	Au	Thorium	Th
Hydrogen	H	Tin (Stannum)	Sn
Iodine	I	Titanium	Ti
Iridium	Ir	Tungsten or Wolfram	W
Iron (Ferrum)	Fe	Vanadium	V
Lantanum	La	Uranium	U
Lead (Plumbum)	Pb	Yttrium	Y
Lithium	Li	Zinc	Zn
Magnesium	Mg	Zirconium	Zr

The foregoing symbols represent, at the same time, the chemical equivalents of the elements. Thus the letters H and O express not hydrogen and oxygen indefinitely, but a single equivalent of these elements: that is, on the hydrogen scale, 1 part of hydrogen and 8 parts of oxygen; and on the oxygen scale, 12·5 parts of hydrogen and 100 parts of oxygen. The symbol of itself expresses only *one* equivalent, and, when several equivalents are to be indicated, this symbol may be repeated: thus, OOO would signify three equivalents of oxygen; or else a figure shewing the number of equivalents may be placed immediately before the symbol, thus, 3O; or, what is preferable, a smaller figure may be placed after the symbol, either above or below it, thus, O^3 or O_3.

Now the association of symbols either with or without the + sign, signifies combination: thus, the formula K + O or KO represents a compound of one equivalent of potassium and one equivalent of oxygen, which is the alkali potash. H + Cl or HCl represents hydrochloric acid, composed of one equivalent of hydrogen and one equivalent of chlorine. SO_3 signifies sulphuric acid, a compound of one equivalent of sulphur with three equivalents of oxygen.

Something more may be expressed in symbolic notation than the names of the elements of a compound body and the numbers of their equivalents; the constitution of the compound, or the mode in which the elements are arranged, may also be represented. The formula, $Fe\,S\,O_4$ expresses truly the composition of proto-sulphate of iron (copperas), but not the arrangement of the elements; on the contrary, the formula $FeO + SO_3$ or FeO, SO_3 signifies that the salt in question is composed of oxide of iron and sulphuric acid, the + sign, or the comma, indicating a distribution of the elements of sulphate of iron into its proximate constituents, oxide of iron and sulphuric acid.

The small figure on the right hand of a symbol does not apply to any other symbol than that to which it is immediately attached; but a large figure placed before the symbol, like a co-efficient in algebra, affects the whole compound expressed, or at least all the symbols before the first comma, or plus sign. Thus $3NO_5$ and $2SO_3$ signify three equivalents of nitric acid and two equivalents of sulphuric acid. The formulæ $2HO, SO_3$ and $2HO + SO_3$ express not twice HO, SO_3 but a compound of two equivalents of water with one equivalent of sulphuric acid; the SO_3 not being affected by the figure 2 at the beginning, because of the interposition of the comma or plus sign. To make the whole symbol subject to the influence of the figure at the beginning, it should be enclosed within a parenthesis, thus $2(HO, SO_3)$ represents two equivalents of hydrated sulphuric acid, and $2(MgO, SO_3 + 7HO)$ two equivalents of crystallized sulphate of magnesia. The following formulæ of some minerals afford examples of the application of these rules:

Felspar, $KO, SiO_3 + Al_2O_3, 3SO_3$.
Analcime, $3NaO, 2SiO_3 + 3Al_2O_3, 2SiO_3 + 6HO$.
Apophyllite, $8(CaO, SiO_3) + KO, 2SiO_3 + 16HO$.

To avoid indistinctness or confusion from the great length of the formulæ of some bodies, several abbreviations have been introduced. When two equivalents of an element are to be expressed, a line is sometimes drawn through the symbol or placed under it; thus, $\not{\mathrm{H}}$ and $\underline{\mathrm{H}}$ signify two equivalents of hydrogen. An equivalent of oxygen in a compound is expressed by a dot placed over the symbol of the other element, the number of dots being the same as the number of equivalents of oxygen; $\dddot{\mathrm{S}}$ expresses sulphuric acid, and $\dot{K}$ potash; therefore $\dot{K}\dddot{S}$ stands for sulphate of potash. Alum is represented by $\dot{K}\dddot{S}, \underline{\dddot{Al}}\,3\dddot{S} + 24\dot{H}$. Sulphur in a compound is sometimes represented in a similar manner by a comma placed over the symbol of the other element, selenium by the sign –, and tellurium by the sign +; but such abbreviations will not be made use of in the present work.

The vegetable and animal acids are conveniently represented by their initial letter with a dash placed over it; thus, $\overline{A}$ stands for acetic acid, $\overline{C}$ for citric acid, $\overline{Tar}$ for tartaric acid, and $\overline{F}$ for formic acid.

In addition to the table of the equivalents of the elements at page 16, the following, which contains the equivalents of several acids and bases, will be found convenient for reference.

TABLE OF EQUIVALENTS.

I. ACIDS.

Acetic (dry), $\bar{A} = (C_4H_3O_3)$	51·0
Arsenic, $As\ O^5$	115·4
Arsenious, $As\ O_3$	99·4
Benzoic, $\overline{Bz} = (C_{14}H_5O_3)$	113·0
Boracic, $B\ O_3$	34·9
Bromic, $Br\ O_5$	118·5
Carbonic, $C\ O_2$	22·0
Chloric, $Cl\ O^5$	75·5
Chromic, $Cr\ O^3$	52·2
Citric, $\bar{C} = (C_4H_2O_4)$	58·0
Formic, $\bar{F} = (C_2HO_3)$	37·0
Gallic, $\bar{G} = (C_7\ H\ O_3)$	67·0
Hydriodic, $H\ I$	127·6
Hydrobromic, $H\ Br$	79·4
Hydrochloric, $H\ Cl$	36·5
Hydrocyanic, $H\ Cy$	27·4
Hydrofluoric, $H\ Fl$	19·7
Hydrosulphuric, $H\ S$	17·0
Hypermanganic, $Mn_2\ O_7$	112·0
Hyposulphuric, $S_2\ O_5$	72·3
Hyposulphurous, $S_2\ O_2$	48·3
Iodic, $I\ O_5$	166·6
Lactic, $\bar{L} = (C_6H_5O_5)$	81·0
Malic, $\bar{M} = (C_8H_4O_8)$	116·0
Manganic, $Mn\ O_3$	51·7
Nitric, $N\ O_5$	54·0
D° hydrated, $H\ O, N\ O_5$	63·0
Oxalic, $C_2\ O_3$	36·0
Phosphoric, $P\ O_5$	71·5
Silicic, $Si\ O_3$	46·3
Sulphuric, $S\ O_3$	40·1
D° hydrated, $H\ O, S\ O_3$	49·1
$2(H\ O, S\ O_3)$	98·2
Sulphurous, $S\ O_2$	32·1
Tannic, $\overline{Tn} = (C_{18}H_5O_9)$	285·0
Tartaric, $\overline{Tar} = (C_8H_4O_{16})$	132·0

II. BASES.

Alumina, $Al_2\ O_3$	51·5
Ammonia, $N\ H_3$	17·0
Antimony, oxide of,	153·0
Barytes, $Ba\ O$	76·7
Chromium, oxide of, $Cr_2\ O_3$	102·4
Cobalt, oxide of, $Co\ O$	37·6
Copper, protox. of, $Cu\ O$	39·7
Copper, suboxide of, $Cu_2\ O$	71·4
Iron, peroxide of, $Fe_2\ O_3$	78·4
Iron, protoxide of, $Fe\ O$	35·2
Lead, protoxide of, $Pb\ O$	111·7
Lime, $Ca\ O$	28·5
Magnesia, $Mg\ O$	20·7
Manganese, protoxide of, $Mn\ O$	37·7
Mercury, oxide of, $Hg\ O$	109·4
Mercury, suboxide of, $Hg_2\ O$	210·9
Nickel, oxide of, $Ni\ O$	37·6
Platinum, oxide of, $Pt\ O$	106·8
Potash, $K\ O$	47·3
Silver, oxide of, $Ag\ O$	116·3
Soda, $Na\ O$	31·3
Strontian, $Sr\ O$	51·9
Tin, peroxide of, $Sn\ O_2$	74·9
Tin, protoxide of, $Sn\ O$	66·9
Zinc, oxide of, $Zn\ O$	40·3

GAS ILLUMINATION.

§ I. History of the Progress of Gas Illumination.—II. Composition of Coal.—III. Hydrocarbons.—IV. Process of making Coal Gas.—V. Secondary products of the Coal Gas Manufacture.—VI. Oil Gas.—VII. Resin Gas.—VIII. Mode of Burning Gas.—IX. Economy of Gas Illumination.—X. Modes of estimating the Illuminating Power and Purity of Light-Gas.—XI. Regulators and Meters. Naphthalized Gas.

Of the numerous services which chemistry has rendered domestic economy, there is certainly none which possesses more general importance and interest, and none that has been more fully or more successfully developed, than that of lighting by means of gas obtained by the destructive distillation of carbonaceous substances.

The process of making gas from such bodies is simple, and an outline of it may be given in a few words.

The material commonly employed as the source of gas is *coal*, that being the most accessible; but any substance of a resinous, fatty, or bituminous nature may also be made use of. Such bodies contain a large proportion of carbon and hydrogen, with, relatively, a small amount of oxygen; the two former elements are the essential constituents of gas for illumination. To make gas from coal, that substance is heated to redness in a large cast-iron tube, or *retort*, placed horizontally in a furnace; the volatile matters which are given off from the coal are conducted by a tube from the retort into a receiver called the *hydraulic main*, in which the liquid products of the distillation are, to a great extent, separated from the gaseous products. The former consist of a black oily liquid known as *coal tar*, and a watery fluid known as the *ammoniacal liquor*. The gaseous product of the distillation, which is a very heterogeneous mixture, and unfit in its present state to be burned within doors, is purified by transmitting it, in the first place, through cold iron tubes, in which it deposits a further quantity of condensable matter; and, in the next place, through milk of lime, or through layers of damp hydrate of lime, to absorb sulphuretted hydrogen and carbonic acid. It is sometimes, lastly, freely washed with water, by which a quantity of ammonia is withdrawn. The gas is then conducted to the gasometer, from whence it is distributed by pipes as may be required.

To procure gas from tar, resin, oils, and fats, these substances are introduced, by small quantities at a time, into a retort, which contains pieces of brick or coke heated to redness; a combustible gas is given off in abundance, and the only purification it requires is cooling, to deposit its more easily condensable constituents.

§ I. HISTORY OF THE PROGRESS OF GAS ILLUMINATION.

Although the application of the gases produced by the destructive distillation of pit-coal as a means of procuring artificial light is of modern invention, yet the germ of it may be traced back nearly two hundred years. In the year 1659, Thomas Shirley is said to have attributed the exhalations from the burning well of Wigan, in Lancashire, to the subjacent coal-beds; and, soon after, Mr. Clayton, rector of Crofton, at Wakefield, in Yorkshire, actually prepared gas by the distillation of coal. In a letter addressed to the Royal Society, May 12, 1688, (*"giving an account of several observations made in Virginia, and in the voyage thither, more particularly concerning the air,"*) speaking of the thunder in Virginia and its effects, Mr. Clayton observes, "I have been told by very serious planters, that thirty or forty years ago, when the country was not so open as it is now, the thunder was more fierce; and that sometimes, after violent thunder and rains, the roads would seem to have perfect coats of brimstone: and it is frequent, after much thunder and lightning, for the air to have a perfect sulphureous smell. Durst I offer my weak reasons when I write to so great masters thereof," (meaning the council of the Society,) "I should here consider the nature of thunder, and compare it with some sulphureous spirits which I have drawn from coals that I could no way condense, yet were inflammable; nay, would burn after they passed through water, and that seemingly fiercer, if they were not overpowered therewith. I have kept of this spirit a considerable time in bladders; and though it appeared as if it were only blown with air, yet, if I let it forth and fired it with a match or candle, it would continue burning until all were spent." Phil. Trans. 1693, No. 201, page 788.*

About the year 1770, a spontaneous evolution of gas took place at a colliery near Whitehaven, belonging to Sir James Lowther. It caught fire on the approach of a lighted candle, and afforded a flame of more than two yards in height. To prevent annoyance to the workmen, it was conducted from the

* The Philosophical Transactions for 1739 contains a letter on the "*spirit of coals,*" from the same author, addressed to the Hon. Robert Boyle, communicated by Dr. Robert Clayton, Bishop of Cork and Orrery. In this he enters more minutely into the products of the distillation of coal. As Boyle died in 1691, it was probably written about the same time as the preceding letter.

"Having seen a ditch within two miles of Wigan, in Lancashire, wherein the water would seemingly burn like brandy, the flame of which was so fierce that several strangers have boiled eggs over it; the people thereabouts affirmed, indeed, that about thirty years ago it would have boiled a piece of beef; and that, whereas much rain formerly made it burn fiercer, now, after rain, it will scarcely burn at all. It was after a long-continued season of rain that I came to see the place and make some experiments, and found, accordingly, that a lighted paper, though it were waved all over the ditch, the water would not take fire. I then hired a person to make a dam in the ditch and fling out the water, in order to try whether the steam which arose out of the ditch would then take fire, but found it would not. I still, however, pursued my experiment, and made him dig deeper; and, when he had dug about the depth of half a yard, we found a shelly coal, and, the candle being then put down into the hole, the air catched fire and continued burning.

"I got some coal and distilled it in a retort in an open fire. At first there came over only phlegm," (ammoniacal liquor,) "afterwards a black oil," (tar,) "and then, likewise, a spirit arose which I could no ways condense; but it forced my lute or broke my glasses. Once when it had pit by a tube which terminated four yards above the surface. The gas issued with such force from the extremity of this tube that large bladders might be filled with it in a few seconds; these were fitted up with little pipes, through which the gas might be pressed out and burned when required. A notice of this evolution of gas, or "blower," as it is now termed, was communicated to the Royal Society in 1773.

About the same time, the production of a permanent gas by the destructive distillation of coal, was noticed by Dr. Hales and by Dr. Watson. The former, by distilling 158 grains of Newcastle coal,

forced my lute, coming close thereto in order to try to repair it, I observed that the spirit which issued out caught fire at the flame of the candle, and continued burning with violence as it issued out in a stream, which I blew out and lighted again alternately for several times. I then had a mind to try if I could save any of this spirit; in order to which I took a turbinated receiver, and putting a candle to the pipe of the receiver, whilst the spirit arose, I observed that it catched flame and continued burning at the end of the pipe, though you could not discern what fed the flame. I then blew it out, and lighted it again several times; after which I fixed a bladder, squeezed and void of air, to the pipe of the receiver. The oil and phlegm descended into the receiver, but the spirit still ascending blew up the bladder. I then filled a good many bladders, and might have filled an inconceivable number more; for the spirit continned to rise for several hours, and filled the bladders almost as fast as a man could have blown them with his mouth, and yet the quantity of coals distilled was inconsiderable. I kept this spirit in the bladders a considerable time, and endeavoured several ways to condense it, but in vain; and, when I had a mind to divert strangers or friends, I have frequently taken one of these bladders and pricked a hole therein with a pin, and compressing gently the bladder near the flame of a candle till it once took fire, it would then continue flaming till all the spirit was compressed out of the bladder; which was the more surprising, because no one could discern any difference in appearance between these bladders and those which are filled with common air." Such observations as these in an age more alive to economic improvements than that in which they were written might have been the means of the general introduction of gas illumination.

obtained 51 grains of incondensable air (*Vegetable Statics*). Dr. Watson observes in his *Chemical Essays*, "I took 96 ounces of Newcastle coal, and, putting them into an earthen retort, distilled them with a fire gradually augmented till nothing more could be obtained from them. During the distillation there was frequent occasion to give vent to an elastic vapour which would otherwise have burst the vessels employed in the operation." When the weights of the liquid products of the distillation and the remaining coke were added together, a loss remained of twenty-eight ounces, which Dr. Watson concluded to be the same kind of elastic vapour as had been previously obtained by Dr. Hales. The inflammability of the gas is also alluded to elsewhere.

It is uncertain to whom the credit of first suggesting the general employment of gas of any kind as a means of procuring artificial light is to be ascribed. It is stated in several French works that the idea of employing gas obtained by the destructive distillation of wood for purposes of illumination originated with Philip Lebon, an engineer, in 1785 or 1786, but Lebon made no communication on the subject to the Institute until the year 1799. In the following year he obtained a patent, and in 1801 published a memoir containing an account of his process under the title, "*Thermolampes ou poêles qui chauffent, éclairent avec économie, et offrent, avec plusieurs produits précieux, une force motrice applicable à toute espèce de machines.*" Lebon states that all unctuous substances afford, by being distilled, a gas fit for illumination: but he does not appear to have made any experiments on the distillation of coal; the material which he generally employed was wood. As the gas from that source possesses very feeble illuminating power, Lebon's speculation, as might have been anticipated, proved a complete failure; he abandoned the enterprise, and established a pyroligneous acid manufactory near Versailles, in which he made the combustible gas disengaged from the wood subservient to the heating of his retorts.

But, while Lebon was making these unsuccessful attempts to introduce gas illumination into France, two individuals were engaged with the same subject in England, with a better promise of success; their source of gas being, not wood, but coal.

The first application of *coal-gas* in illumination appears to have been made in 1792 by Mr. William Murdoch, engineer to Messrs. Bolton and Watt, and then residing at Redruth in Cornwall; but the application did not extend beyond his own dwelling-house and offices. Mr. Murdoch removed to Ayrshire a few years afterwards, where he erected another gas apparatus in 1797. The first application of gas on the large scale was made in the following year, when Mr. Murdoch fitted up a gas-work at the manufactory of Messrs. Bolton and Watt at Soho.

Novel as was this mode of illumination, it did not attract the notice of the public until the year 1802, when the front of the Soho manufactory was lighted by Mr. Murdoch with a public display of gas-lights, on occasion of the national illumination in the spring of that year at the peace of Amiens.

The superiority of the new light over the dim oil-lights in use at that day excited no small degree of popular attention. An immense crowd came to view the spectacle, and an account of it was circulated throughout the country in the public papers. Gas-works were soon afterwards erected in Birmingham, Manchester, Halifax, and other towns. In 1804 and 1805 the extensive cotton-mill of Messrs. Philips and Lee, at Manchester, was lighted with gas under the superintendence of Mr. Murdoch. In this establishment were nine hundred burners, producing a light equal to that of two thousand five hundred candles; the quantity of gas made amounted on the average to twelve hundred and fifty cubic feet per hour. Mr. Murdoch sent a detailed account of his operations to the Royal Society in 1808, for which he received a gold medal. (*An account of the application of gas from coal to economical purposes*, Phil. Trans. 1808, p. 124.)

But although gas illumination was extensively adopted in manufactories, it made very little progress in London, probably on account, in some measure, of the insufficiency of the means then known of effecting its purification. In manufactories, where the ventilation is generally good, the inconvenience arising from the disagreeable odour of impure gas was comparatively of little consequence; but in shops, private houses, churches, &c. the impurity of the gas would be a serious impediment to its general adoption.

The indefatigable exertions, however, of Mr. Winsor in London, at the same time that Mr. Murdoch was carrying on his operations in the country, were very efficient in introducing gas illumination into the metropolis. In 1803 and 1804 the Lyceum theatre, and soon afterwards one side of Pall Mall, were lighted by Mr. Winsor with coal-gas, the manufacture of which soon suggested itself to the public as a lucrative speculation. In 1804 Mr. Winsor obtained a patent for some apparatus employed in the process, and established the National Light and Heat Company for carrying on the manufacture on an extensive scale. In 1809 this company applied to parliament for a charter, but they were opposed by Mr. Murdoch on the score of priority, and the charter was refused. On a subsequent application, however, the claims of Mr. Winsor were recognised, and the charter was granted. In the course of from ten to fifteen years the oil-lamps of every street and alley of the metropolis were displaced by gas, churches and other public buildings and shops were illuminated by the same light, and gas has now become general in almost every town in the empire.

For many improvements in the construction of the apparatus, and in the practical details of the process; gas illumination is indebted to Mr. Samuel Clegg, whose attention was called to this subject so early as the year 1804, and who erected a gas-work of pretty large extent in the following year. Among the improvements introduced by Mr. Clegg may be mentioned the hydraulic main, by which the retorts are isolated while being emptied and charged; the wet lime purifier, and the meter.

During a parliamentary investigation, in 1823, it

was ascertained that the Chartered Gas Company (the original National Light and Heat Company) alone consumed annually at their three stations (Peter Street, Westminster; Brick Lane, St. Luke's; and Curtain Road, Shoreditch,) 20,678 chaldrons of coals, and produced on the average 680,000 cubic feet of gas every night. This quantity supplied more than thirty thousand burners, affording a light equal to as many pounds of tallow-candles. At the Peter Street station the number of retorts then erected amounted to three hundred, and of gasometers fifteen; the extent of the main pipes belonging to this establishment for distributing the gas was fifty-seven miles. The number of retorts at the Brick Lane works in 1822 was three hundred and seventy-one, of which one hundred and thirty-three were worked on the average of summer and winter; the number of gasometers was twelve, and the length of the distribution pipes forty miles. At the Curtain Road works the number of retorts was two hundred and forty, but the greatest number worked in 1821 was eighty; the number of gasometers was six, and the length of the main twenty-five miles.

At the same time three other large gas companies were in existence, namely, the City of London Gas-light Company, Dorset Street; the South London Gas-light and Coke Company, with two stations, one at Bankside and the other at Wellington Street; and the Imperial Gas-light and Coke Company, whose establishment at Hackney was then in the course of erection. The extent of the entire operations of these companies was a little more than equal to that of the Chartered Company.

Since the time at which the above estimates were made, several other companies have been established in London, and the amount of gas made in the various works has been increased at least seven-fold. The Chartered Company's works, which are supposed to supply about a fifth part of all the gas burned in London and the suburbs, consume no less than fifty thousand chaldrons of coals annually; and, taking the average product of purified gas from a chaldron of coals at twelve thousand cubic feet, the amount annually made by the Chartered Company is six hundred millions of cubic feet, or, expressed by weight, more than eighteen million pounds. The gas-works next in extent to the Chartered Company's is that of the London Gas-light Company, whose establishment at Vauxhall is, probably, the most complete in arrangement, and the most powerful of any in the world. The extent of their mains, which ramify into Middlesex as well as Surrey, exceeds one hundred and fifty miles; and, by the power of the works and the admirable mode in which the pipes are laid, gas may be supplied to a place more than seven miles distant, in the same quantity and with the same precision as at Vauxhall.

The entire annual consumption of gas in London and the environs is estimated at not less than three thousand millions of cubic feet, and the light produced by the combustion may be considered as equal to what would be obtained from one hundred and sixty millions of pounds of tallow-candles. The annual consumption of coals in the various London gas-works is said to be about two hundred and fifty thousand chaldrons, and nearly nine hundred tons per day are consumed in foggy weather in winter.

Although the French tenaciously assert the claim of their countryman, Lebon, to the merit of having first applied gas as a source of light, yet this mode of illumination was not adopted in France until the year 1818, when M. Chabrol de Volvic, then prefect of the Seine, constructed, at the Hospital of Saint Louis at Paris, an apparatus which supplied gas to fifteen hundred burners. Since that time four other gas-works have been erected at Paris. The provincial French towns which possessed gas-works in 1839 were, Lyons, Reims, Amiens, Havre de Grace, and Elbeuf. Refuse oily matters are now common as sources of gas on the continent.

Notwithstanding the present enormous production and consumption of gas, there can be no doubt that gas illumination is far from having attained that developement of which it is susceptible. Gas is still an article of increasing consumption; and in proportion as attention is paid to improvements in its manufacture, in the fittings, and in the mode of burning, and to the means of conducting away the products of the combustion, it will become more and more generally adopted in private dwellings, so as to supersede at last almost all other sources of artificial light.

Before entering on the minute details of the process of making illuminating gases, it will be convenient to consider briefly, in the first place, the composition of different kinds of coal as the ordinary sources of gas, and, in the next place, the composition and properties of those hydrocarbons which form the most important constituents of light-gas.

§ II. COMPOSITION OF COAL.

The difference in the appearance of the several varieties of coal is not greater than the difference in their composition. The principal constituents of coal are carbon, hydrogen, oxygen, and nitrogen, the first three of which may be considered as the elements of *pure* coal; but it always contains likewise a great number of earthy impurities. The different kinds of coal comport themselves in the fire as ordinary organic substances in which the combustible elements, carbon and hydrogen, are condensed into a very small volume. A very inferior coal contains a much larger proportion of carbon and hydrogen than wood.

An insight into the composition of coal, sufficient to determine the relative value of the different varieties in most of their practical applications, may be obtained by determining, 1st, the quantity of volatile matters given off by the coal when heated to bright redness without access of air; and, 2ndly, the proportions which exist between the carbon and the ash, or earthy matter, in the remaining coke, which may be ascertained by incineration in the open air. For gas illumination, the value of the coal is proportional to the quantity of volatile matters which it disengages at a red heat. The following are the results of examinations of this kind on a few varieties of coal, by Mr. Mushet.

	Volatile Matter.	*Coke.*	
		Carbon.	*Ash.*
Scotch cannel coal	56·570	39·430	4·000
Derbyshire ditto	47·000	48·362	4·638
Welsh furnace coal	8·500	88·068	3·432
Welsh stone coal	8·000	89·700	2·300
Kilkenny coal	4·250	92·877	2·873

The most accurate method of determining the composition of coal is by an ultimate analysis, after the manner of an ordinary organic analysis. The following table contains the results of ultimate analyses of several varieties of coal by different chemists:

	Carbon.	*Hydrogen.*	*Oxygen.*	*Nitrogen.*
1. Caking coal	75·28	4·18	4·58	15·96
2. Splint coal	75·00	6·25	12·50	6·25
3. Cherry coal	74·45	12·40	2·93	10·22
4. Cannel coal	64·72	21·40	——	13·72
5. Ditto	72·22	3·93	21·05	2·08
6. Ditto	74·83	5·45	19·72	
7. Ditto	70·90	4·30	24·80	
8. Newcastle coal	84·99	3·23	11·78	

The analyses Nos. 1, 2, 3, and 4 of this table are by Dr. Thomson; No. 5, by Dr. Ure; Nos. 6 and 8, by M. Karsten; and No. 7, by Mr. Crum.

As most, if not all kinds of coal, give, on distillation, a larger or smaller proportion of ammoniacal products, the existence of nitrogen is probably constant, though not indicated by the analyses of Crum and Karsten. The quantity of that element, however, may not exceed a mere trace in some kinds of coal. The proportion of nitrogen obtained by Dr. Thomson seems to be in very considerable excess.

The composition of *pure* splint and cannel coal, according to analyses by Richardson and Regnault, is represented by the formula $C_{24} H_{13} O$; which gives, when expressed in proportions per cent.

Carbon	87·27
Hydrogen	7·88
Oxygen	4·85
	100·00

Pure caking coal, according to Liebig, has the composition $C_{20} H_9 O$, or, per cent., carbon, 87·59; hydrogen, 6·57; and oxygen, 5·84. Nitrogen does not seem to be an element of pure coal any more than of pure lignin or woody fibre, from which the coal itself is derived. The form in which the nitrogen exists is uncertain; though given off as ammonia on the destructive distillation of the coal, it certainly does not pre-exist in that form, with the exception of a mere trace. It is probably derived, in part, from the albuminous matter contained in the original wood, but it may also be traced to the animal remains found in the coal.

Some kinds of coal contain a small quantity of a resinous matter, which may be extracted by digesting the coal in ether, or oil of turpentine: the resin dissolves, and leaves the other constituents of the coal unacted on. Such coal is probably derived from the decomposition of trees which contain a considerable quantity of essential oil and resinous matter, such as the pine, for instance.

The extraneous constituents of the commoner varieties of coal are the following:

Argillaceous matter, derived from the contiguous

beds, is the earthy impurity usually found in greatest quantity. It is composed chiefly of silica and alumina, but may also contain magnesia, oxide of iron, and potash; the latter is derived from the decomposition of felspar (see page 16). It is mixed so intimately with the coal as to be invisible by the naked eye, and can only be recognised (except by analysis) by the hardness and tenacity which it communicates to the coal, especially when in a considerable proportion.

Carbonate of iron is very frequently, if not always, found in coal. It is not disseminated minutely, like the argillaceous matter; but, for the most part, in detached reniform masses of various sizes.

Unfortunately for some applications of coal, and especially that of making gas, few varieties are free from *iron pyrites*, or *bisulphuret of iron*. This mineral is generally visible by the naked eye; it exists disseminated in minute crystals and in layers in fissures of the coal. The sulphuretted hydrogen contained in impure coal-gas is derived almost entirely from iron pyrites.

Carbonate of lime is generally present in small quantity, sometimes minutely disseminated, like the argillaceous matter, but often in thick laminæ.

A constant product of the distillation of coal is *muriate of ammonia*, which shews that chlorine must exist in the coal in some shape or other. *Chloride of sodium* (common salt), which is contained in the water infiltrated into the coal-mines, may be the source of the chlorine; but muriate of ammonia itself has been detected by M. Bussy in a specimen of coal from Commentry, near Montluçon. A trace of *hydriodate of ammonia* was also found by M. Bussy.

Sulphate of lime (gypsum) is considered by Lampadius to be a constant impurity in coal. Of all the specimens he examined, comprising more than thirty in number, not one was found to be free from that substance (*Manuel de Métallurgie*).

Besides the impurities mentioned, coal has sometimes been found to contain *quartz*, *mica*, *sulphate of barytes*, *galena*, *blende*, *dolomite*, *phosphate of lime* (in reniform masses, like carbonate of iron), *sulphate of alumina*, *sulphate of iron*, traces of *manganese*, *magnesia*, *sulphur* in the free state (M. Bussy), and a *silicate of alumina* of a peculiar composition (named *pholerite* by M. Guillemin). *Silicate of soda* has been detected in the water which infiltrated through coal-beds (Mr. Leigh).

Spurious coal, (called, in Scotland, *parrot coal*,) which is found among strata of genuine coal, contains about one-fifth of its weight of stony matter. *Anthracite* leaves, when burned, nearly 40 per cent. of ash.

So far as the use of coal in making gas is concerned, the impurity of greatest moment is iron pyrites—the principal source of sulphuretted hydrogen. In the extensive purification required by the gas when the quantity of sulphuretted hydrogen is considerable, a notable proportion of useful gas, or vapour, is separated, together with the sulphuretted hydrogen. Hence the advantage of ascertaining, by analysis, the relative proportions of sulphur in the different varieties of coal, in order to select that which contains the smallest quantity. It has been estimated that, from coal nearly free from sulphur, gas may be obtained which affords ten per cent. more light than that from the common sulphurous coal. It is not likely that any variety of coal would be found entirely free from a source of sulphuretted hydrogen, but the proportions of iron pyrites vary considerably in the different kinds.

§ III. HYDROCARBONS.

The substances to which the illuminating power of coal-gas is due, are compounds of carbon and hydrogen, and are therefore termed hydrocarbons. These bodies constitute an extensive class of substances, including solids, liquids, and gases, possessing, with a great diversity of physical characters, a remarkable similarity in composition. As a class, these bodies possess considerable interest on account of the useful applications which some of them have received in domestic economy, and of their production in various natural and artificial processes. These compounds have an organic origin, being either formed under the influence of a vital process, or derived from the decomposition of substances containing carbon and hydrogen, which once formed part of an organized body.

It is certainly a remarkable fact that bodies differing from each other in appearance so greatly as essential oil of lemons, coal-gas, spirit of turpentine, naphtha or petroleum, bitumen, attar of roses, naphthalin, and caoutchouc, should be composed of the same two elements, though united in different proportions; but it has been shewn that some of these hydrocarbons actually possess the same ultimate composition, or proportions per cent. of the two elements, though unquestionably different substances. Essential oil of lemons and oil of turpentine, for instance, are both composed of

Carbon	88·2
Hydrogen	11·8
	100·0

The term *isomerism* (from ισος, equal, and μερος, part,) is made use of to express such a relation in composition between different bodies, and the bodies themselves are said to be *isomeric*. When the doctrine of isomerism was first introduced, it was believed that isomeric bodies may have not only the same ultimate composition, but the same constitution (see page 31), or the same arrangement of the ultimate atoms. But more recent research has been unfavourable to such a view, and the term isomerism is now very commonly and generally employed to indicate identity in the composition of two or more bodies when expressed in one hundred parts, not necessarily accompanied by a similar arrangement, or number of the ultimate atoms. In the example above cited, the proportion per cent. of carbon and hydrogen shews the relative proportion of the equivalents to be five equivalents of carbon to four equivalents of hydrogen. Now, a closer investigation of the essence of lemons has led to the conclusion that one integrant molecule, or equivalent of that essence, contains ten equivalents of carbon and eight equivalents of hydrogen; while the molecule of oil of turpentine is believed to contain twenty equivalents of carbon and sixteen equivalents of hydrogen. Where such difference in constitution exists, the possession of the same ultimate composition need not be necessarily accompanied by the possession of identical properties.

Several gases and vapours among the hydrocarbons are found to possess exactly the same composition, taking equal *weights;* but, in equal *volumes*, the amount of carbon and hydrogen in the gases, or vapours, is found to be very different, and in multiple proportions. Olefiant gas, for example, has exactly the same composition as a gas obtained by the destructive distillation of oil, taking equal weights; but in equal bulks there is found exactly twice as much matter in the latter as in the former. The composition of olefiant gas is expressed by the formula $C_4 H_4$, and that of oil-gas by $C_8 H_8$. As regards their composition, therefore, these gases differ from each other only in their state of condensation.

The hydrocarbons on which the illuminating power of coal-gas depends are *light carburetted hydrogen* and *olefiant gas.*

Light carburetted hydrogen.—This gas consists, in 100 parts by weight, of 75 of carbon and 25 of hydrogen, proportions which correspond to one equivalent of carbon and two equivalents of hydrogen. Its density is 560·45, compared with air as 1000; and in one volume of it are contained two volumes of hydrogen gas and one volume of carbon vapour. It is colourless, tasteless, quite neutral, and nearly inodorous. It is unable to support the respiration of animals, but is respirable if mixed with air. If a burning taper is introduced into a jar of the gas, the flame of the taper is extinguished; but the gas itself takes fire when in contact with air, and burns with a yellow flame of little intensity. Mixed with a proper proportion of atmospheric air or oxygen, light carburetted hydrogen forms a mixture which explodes violently by the contact of flame, or by the electric spark. It requires twice its bulk of oxygen for complete combustion, and affords water and its own bulk of carbonic acid. It is a compound of considerable stability; is not affected by chlorine gas in the direct rays of the sun, if all moisture is absent; and is only partially resolved into its elements when passed through a tube heated to whiteness. A small quantity of carbon is then deposited within the tube.

Light carburetted hydrogen is a product of the destructive distillation of most organic matters, but it cannot be prepared in a state of purity by such a process. It is a constant product of the putrefactive decomposition of vegetable matter under the surface of water, and may be readily obtained by agitating the mud at the bottom of stagnant pools and collecting the gas as it bubbles up in an inverted bottle and funnel. It is hence sometimes termed the inflammable air of marshes, or marsh gas. As thus obtained, it contains carbonic acid gas, and a small proportion of nitrogen gas. The former may be removed by agitating a little solution of potash in the gas, but no easy method is known of separating the nitrogen.

The best mode of preparing pure light carburetted

hydrogen artificially is, by distilling a mixture of acetate of potash with an alkali at a red heat, in a coated glass or an earthen retort. The gaseous mixture which is obtained on passing the vapour of alcohol through an ignited porcelain tube consists chiefly of this gas.

Although light carburetted hydrogen is instantly kindled by flame if in contact with the air, yet it requires a higher temperature for its inflammation than most other combustible gases. A glass rod, heated to dull redness, is sufficiently hot to ignite hydrogen, sulphuretted hydrogen, carbonic oxide, and olefiant gases; but requires to be heated to bright redness, or whiteness, to set fire to light carburetted hydrogen.

Olefiant gas.—Like light carburetted hydrogen, olefiant gas has not yet been reduced to the solid or liquid state. This gas exists ready-formed in nature, and may be produced artificially by the destructive distillation of fatty, oily, and bituminous matters, and, in fact, the most part of organic substances. It was first recognised as a peculiar body in 1796, by the associated Dutch chemists Bondt, Dieman, Van Troostwick, and Lawerenburg, who gave it its present name from its property of forming an oily-looking compound with chlorine.

This gas is commonly obtained by heating together in a retort a mixture of one part of alcohol with six or seven parts, by weight, of oil of vitriol. The retort should be furnished with a globular receiver, from which an exit-tube leads into the pneumatic trough where the gas is to be collected. The gas which is given off is a mixture of equal volumes of olefiant gas and sulphurous acid, and contains, besides, vapours of water, ether and alcohol. It may be obtained pure by passing it first through milk of lime to retain the sulphurous acid, and afterwards through oil of vitriol to absorb the vapours.

Olefiant gas is a little lighter than air, its specific gravity being 996·87, compared with air as 1000. It is tasteless, but has a peculiar feeble ethereal odour. Like light carburetted hydrogen, it cannot support respiration or combustion. It inflames at a red heat when in contact with the air, and burns with a white and remarkably luminous flame. If mixed with air and exploded, the detonation is extremely violent. Olefiant gas consists, by weight, in 100 parts, of

Carbon	85·71
Hydrogen	14·29
	100·00

or, by volume, of two volumes of carbon vapour and two volumes of hydrogen gas condensed into one volume. Its symbol is C_4H_4. The peculiar luminosity of the flame of this gas, and the force of its explosion when mixed with air or oxygen, are clearly accounted for in its state of condensation.

Olefiant gas requires three times its volume of oxygen for complete combustion, and produces water and two volumes of carbonic acid gas. When exposed to a red heat, as by being passed and repassed through a porcelain tube, it is decomposed into its elements, solid carbon being deposited within the tube, and hydrogen gas liberated, having twice the original volume of the olefiant gas. If a mixture of one volume of olefiant gas with two volumes of chlorine gas contained in a tall and narrow glass jar is inflamed by contact with a lighted taper, the olefiant gas is decomposed, and combustion takes place in a very striking manner: a deep red flame gradually descends through the mixture, resulting from the combination of the chlorine with the hydrogen of the olefiant gas, by which muriatic acid is produced, while the carbon is at the same time deposited as a dense black cloud. If olefiant gas and chlorine gas are merely mixed in equal volumes in a glass jar over water, the two gases gradually combine, and condense into an oily-looking liquid known as oil of olefiant gas, or Dutch liquid. Olefiant gas also unites with iodine, and with bromine. The combination with iodine is a white crystalline solid, which may be sublimed without decomposition in an atmosphere of olefiant gas, but not in the air.

The inflammable gas which escapes from fissures in the earth, before referred to (page 38), known as *fire-damp*, is essentially a mixture of light carburetted hydrogen, free hydrogen, and olefiant gas, differing little in properties and composition from purified coal-gas. It generally proceeds from subterranean deposits of coal, and is found in the working of coal-mines pent up in cavities. It seems to be given off spontaneously from the fresh surface of some kinds of coal in small but sensible quantities. When the gas accumulates in the gallery of a coal-mine where the air is still, so as to form one-seventh or one-eighth part of the atmosphere, the approach of a naked flame instantly causes the mixture to ignite with a most tremendous explosion, by which, before the introduction of the safety-lamp of Sir Humphrey Davy, the life of the coal-miner was placed in the most imminent danger.

An explosive mixture of fire-damp and air is not ignited by iron or charcoal at an incipient red heat, although the contact of *flame* of any kind instantly determines its inflammation. On this circumstance depends the efficacy of the safety-lamp, which is simply a common oil lamp with a chimney of wire gauze, so constructed that all the air which obtains access to the flame is obliged to pass through the gauze. The cooling power of gauze containing about eight hundred apertures in a square inch is sufficient to prevent, under ordinary circumstances, the passage of flame from the interior to the exterior of the lamp. A mixture of fire-damp and air does not explode violently when the quantity of air is much above or below that required for the complete combustion of the gas, which is about seven or eight times its volume. If fire-damp consisted only of light carburetted hydrogen, it should require very nearly ten times its volume of air. A mixture of one volume of light carburetted hydrogen and four of air does not explode, nor does a mixture of one volume of gas and fifteen of air; in the latter case the gas burns about the flame of the taper, but the large quantity of air prevents the temperature of the mixture rising sufficiently high for ignition.

Naphthalin and *Paranaphthalin.*—Light carburetted hydrogen and olefiant gas are the only hydrocarbons which are present in considerable quantity in coal-*gas*, but the volatile *solid* and *liquid* products of the distillation of coal also contain several hydrocarbons, one of the most important of which is *naphthalin.* This substance is a white crystalline solid, sometimes found condensed in certain parts of the coal-gas apparatus in a tolerably pure state, but is most abundantly contained in the tar. It may be obtained by subjecting coal-tar to distillation, the receiver being changed at different periods of the process. The first product which distils over is a yellowish oily-looking liquid known as *coal-oil*, which is a mixture of several distinct substances, some of which are acids and some bases, but little is known respecting their composition. If coal-oil is redistilled and the last products received apart, they afford a considerable quantity of naphthalin when cooled artificially. The naphthalin thus obtained may be purified by solution in hot alcohol and recrystallization. It is stated by M. Laurent that this substance is obtained more abundantly from old than from recent coal-tar; and also that the product is much increased if the coal-oil is subjected for some time to a current of chlorine gas.

When pure, naphthalin has the form of transparent, colourless and shining plates, possessed of a peculiar odour and a burning taste. It fuses at about 176° Fahr., boils at 423° Fahr., and may be condensed without change. It is insoluble in water, but very soluble in alcohol and ether, and is precipitated in a crystalline state from its solutions in these liquids on the addition of water. Naphthalin is readily inflamed, and burns with a strong light and production of much smoke. It is not decomposed by heat into free carbon and hydrogen so readily as most other hydrocarbons, and seems, in fact, to be peculiarly the product of a high temperature; it being produced when wood and most vegetable matters are exposed to the destructive distillation at a bright red heat, but not in so large a proportion at a lower temperature. The composition of naphthalin is the following:

	Atomic weight.	Per-centage composition.
20 eq. carbon	120	93·75
8 eq. hydrogen	8	6·25
1 eq. naphthalin	128	100·00

With the assistance of heat, naphthalin combines with oil of vitriol and with anhydrous sulphuric acid, forming a fine deep purple-red liquid. When this liquid is diluted with water, a considerable proportion of the naphthalin is precipitated in an altered state, and the aqueous solution contains three peculiar combinations of naphthalin (or modifications of that substance) with hyposulphuric acid ($S_2\,O_5$), each of which possesses the properties of an acid. A great number of interesting compounds have also been obtained by the action of chlorine and nitric acid on naphthalin, a short account of which may be found in the last edition of the "Elements of Chemistry" of the late Dr. Turner, and in the "Elements of Chemistry" of Professor Graham.

The name of paranaphthalin has been applied by M. Dumas to a hydrocarbon much resembling naphthalin in appearance, which substance it accompanies in coal-tar. It has the same ultimate composition as naphthalin, but its constitution is considered to be different, its symbol being $C_{30}\,H_{12}$, instead of $C_{20}\,H_8$. Several circumstances serve to distinguish paranaphthalin from naphthalin. The former is insoluble in hot alcohol; it fuses at 356° Fahr., and boils at 572° or higher. If equal weights of paranaphthalin and naphthalin are converted into vapour at the same temperature, it is found that the vapour of paranaphthalin occupies exactly one-third less space than the vapour of naphthalin. The density of the vapour of paranaphthalin is 6630 (air at the same temperature being 1000); the density of the vapour of naphthalin is 4420, or exactly one-third less.

Some other hydrocarbons are given off in small quantity on the destructive distillation of coal, but their composition and properties are not well known. They are mostly liquids, and contained in the tar. A few ternary combinations of carbon, hydrogen, and oxygen, (the last element in small relative proportion,) are also produced in small quantity, some of which are also contained in the tar obtained by the destructive distillation of wood. As the remaining secondary products obtained in the making of coal-gas are not peculiar to that process, a separate description of them here is unnecessary. We proceed, therefore, in the next place, to the process itself.

§ IV. PROCESS OF MAKING COAL-GAS.

By reverting to the introductory observations at page 36, it will be observed that a coal-gas apparatus must be so constructed as not only to generate but purify the gas: it must also have the means of discharging the gas steadily into the pipes for distribution, in order to ensure an uniform exit from the burners. The principal parts of the apparatus are the retort, hydraulic main, cooler, purifier, and gasometer. The design on the following page shews the arrangement of the apparatus when looking at the front of the furnace, or the anterior ends of the retorts:—*a* is the furnace, with a set of five retorts, from each of which a tube proceeds upwards perpendicularly to a height of from ten to twelve feet; it then takes a curve downwards, and enters the hydraulic main *b*. The latter is an iron tube of from twelve to fourteen inches in diameter, placed horizontally: the extremities of the tubes from the retorts dip about two inches below the surface of the liquid in the hydraulic main. From one end of this tube, a little above the centre, proceeds the tube *c*, into a close cistern *d*, in which the tar and ammoniacal liquor are received and separated. From the top of the opposite end of the hydraulic main proceeds the tube *e*, by which the gas is conducted into the cooler *f*, and having deposited here a further quantity of condensable matter, the gas is passed on to the lime purifier *g*, and from thence, commonly, to the gasometer *h*, but sometimes through other apparatus, not represented in the annexed figure, but which will be described afterwards. The construction of each of the parts of a coal-gas apparatus admits of certain modifications, and requires to be considered minutely.

The tabular view of the products of the distillation of coal in the page following the figure will assist in shewing the uses of certain parts of the apparatus.

Fig. 1.

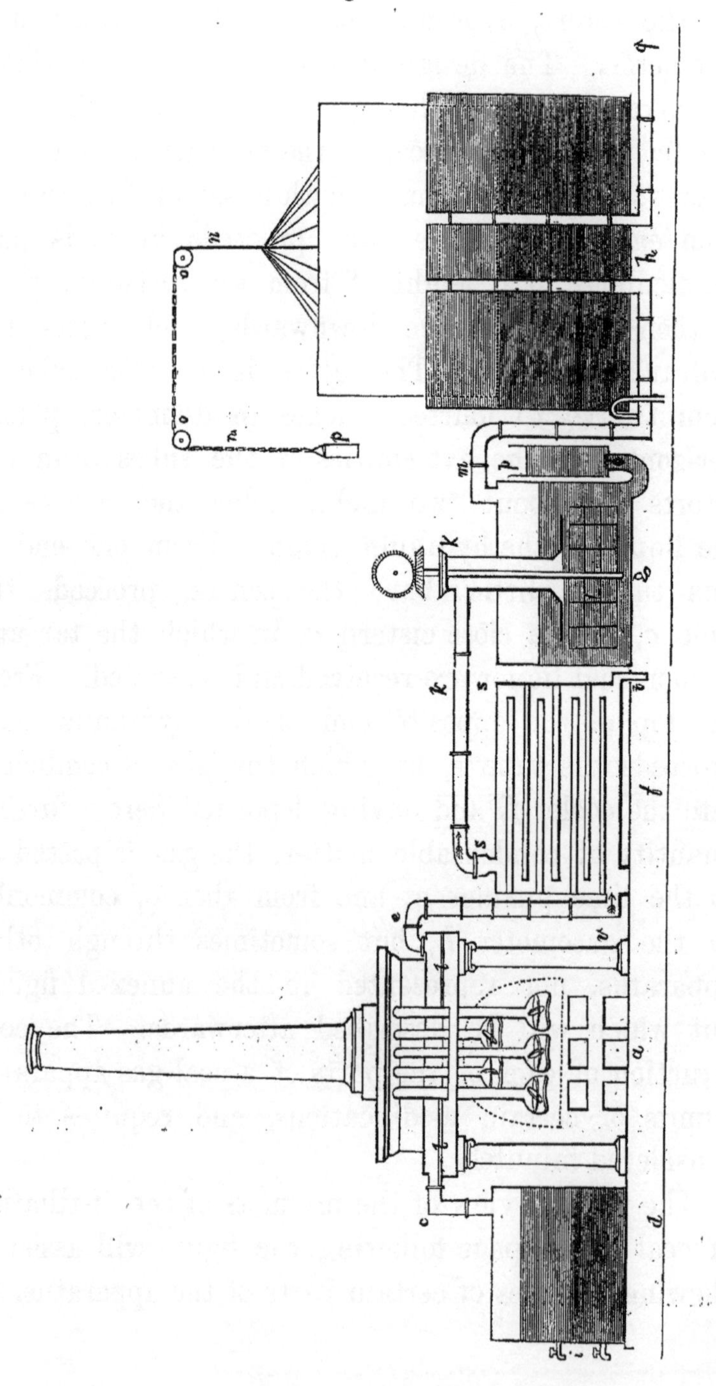

PRODUCTS OF THE DESTRUCTIVE DISTILLATION OF COAL.

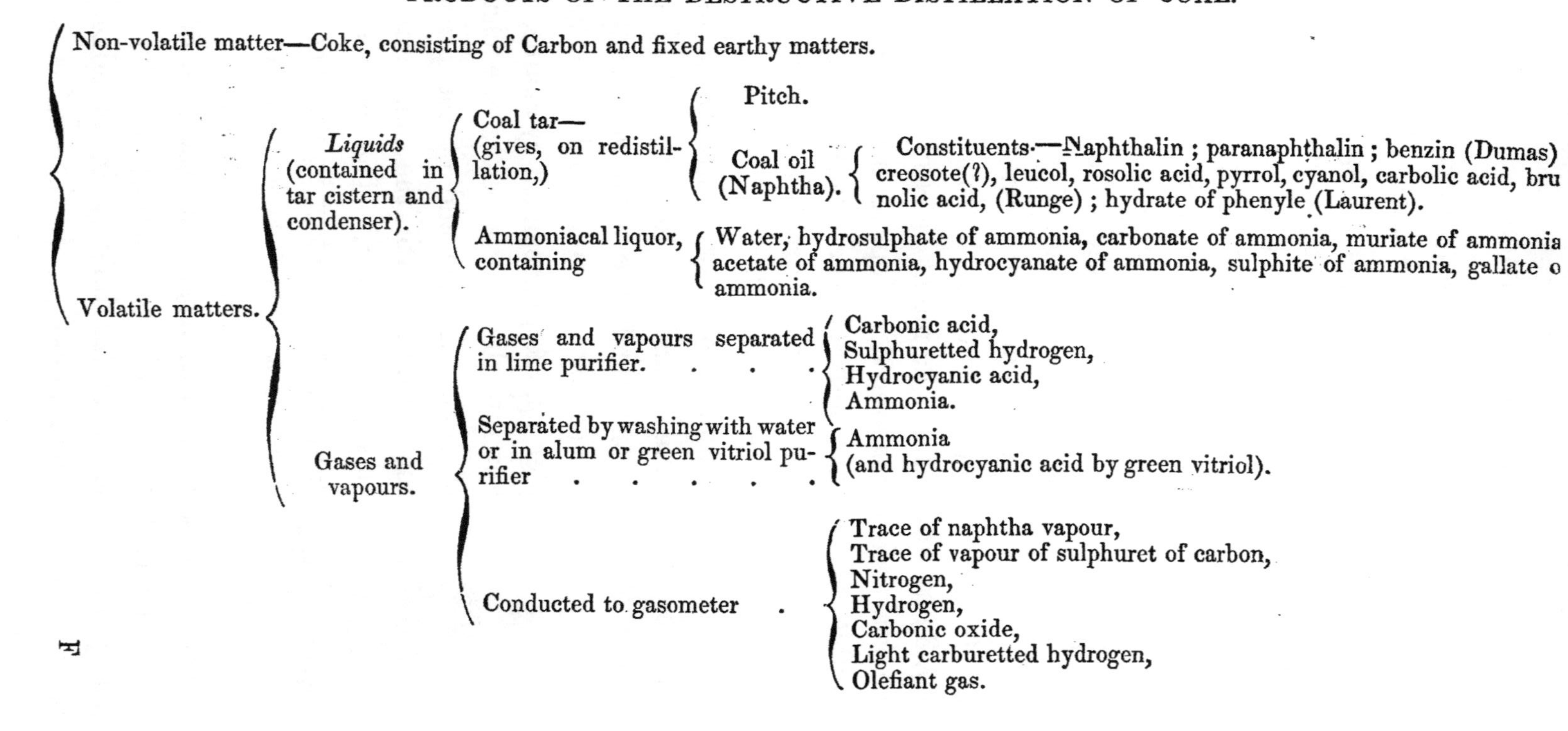

- Non-volatile matter—Coke, consisting of Carbon and fixed earthy matters.
- Volatile matters.
  - *Liquids* (contained in tar cistern and condenser).
    - Coal tar—(gives, on redistillation,)
      - Pitch.
      - Coal oil (Naphtha).
        - Constituents.—Naphthalin ; paranaphthalin ; benzin (Dumas) creosote(?), leucol, rosolic acid, pyrrol, cyanol, carbolic acid, bru nolic acid, (Runge) ; hydrate of phenyle (Laurent).
    - Ammoniacal liquor, containing
      - Water, hydrosulphate of ammonia, carbonate of ammonia, muriate of ammonia acetate of ammonia, hydrocyanate of ammonia, sulphite of ammonia, gallate o ammonia.
  - Gases and vapours.
    - Gases and vapours separated in lime purifier. . . .
      - Carbonic acid,
      - Sulphuretted hydrogen,
      - Hydrocyanic acid,
      - Ammonia.
    - Separated by washing with water or in alum or green vitriol purifier . . . . . .
      - Ammonia
      - (and hydrocyanic acid by green vitriol).
    - Conducted to gasometer .
      - Trace of naphtha vapour,
      - Trace of vapour of sulphuret of carbon,
      - Nitrogen,
      - Hydrogen,
      - Carbonic oxide,
      - Light carburetted hydrogen,
      - Olefiant gas.

The quality of the gas depends in no small degree on the shape and size of the retort. The original form was that of a cylinder placed upright (fig. 2), large enough to hold about fifteen pounds of coal, with the exit-tube at top by the side. Such a retort

Fig. 2. Fig. 3.

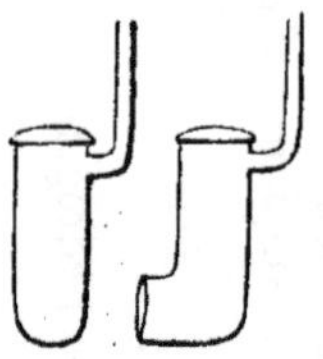

is very readily charged, and the gas produced from it is of good quality; but its form is very inconvenient for the removal of the coke at the end of the process. One of the means devised in order to remedy this defect was to have an opening at the bottom of the retort (fig. 3) at which the coke might be easily withdrawn, the coal being introduced at top, as before; but the trouble of closing both of the apertures at each charge was found to be as great as the removal of the coke in the original form. A much larger retort, of the original form, was afterwards used, capable of holding from ten to fifteen hundred-weight of coal, from which the coke was easily re-

Fig. 4.

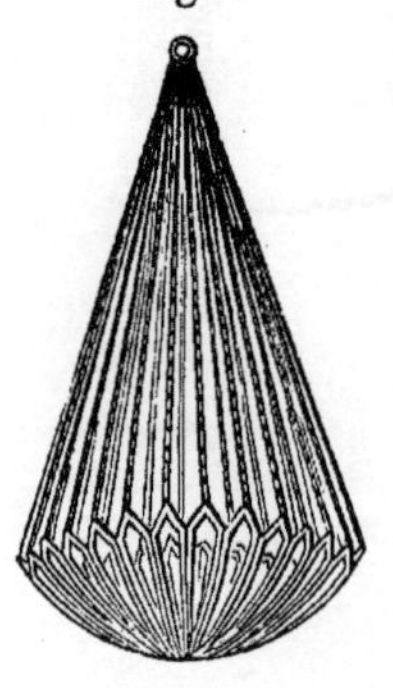

moved by an iron basket or grappler (fig. 4), suspended by chains, and put into the retorts before the charge of coal. At the end of the process, the grappler with the coke was removed by a crane. The great disadvantage of this retort is, that the heat requires a considerable time to penetrate to the interior of the great mass of coal, the cake of coke formed at the side of the retort being a very badly conducting medium, and gas produced slowly is far inferior in illuminating power to that formed quickly, as in the original small retort, for reasons which will afterwards appear. To the latter it was found expedient to return, and the inconvenience experienced in the removal of the coke was lessened by placing the retort in a horizontal position; the coal is usually introduced by a common square spade, and sometimes by a tray of sheet-iron, similar to a grocer's scoop, which is pushed to the end of the retort, inverted so as to turn out the coal, and withdrawn.

The forms of the retort generally used at present are shewn in vertical sections, parallel to the door, at figs. 5, 6, and 7; fig. 5 is called the D or semicircular retort, fig. 6 the kidney-shaped retort, and fig. 7 the elliptical retort. The greatest breadth of each form is about twenty or twenty-two inches, and the height in the middle from nine to twelve inches.

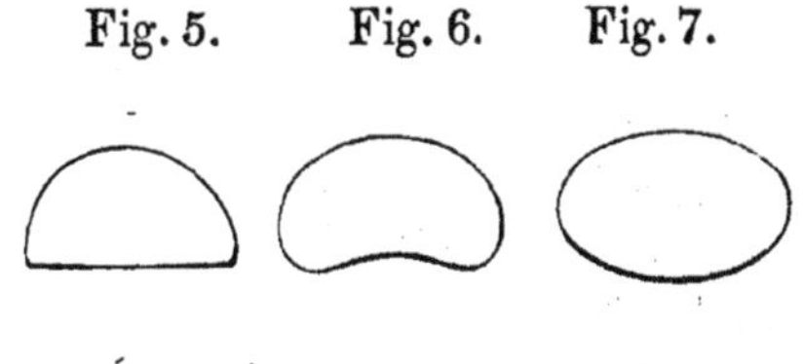
Fig. 5. Fig. 6. Fig. 7.

The kidney-shaped retort is said to require the least heat, and to yield the largest quantity of good gas in the shortest time: the circular or cylindrical retort is seldom employed in well-conducted gas-works, unless its diameter is made less than twelve or fourteen inches.* The length of the retort is commonly seven feet six inches. Cast-iron is the only material used for gas retorts in this country, but in some parts of the Continent baked clay is employed. The ordinary duration of a retort charged four times in twenty-four hours is three months.

The door of the retort is secured by a cross-bar

* The circular retort of ten or eleven inches in diameter is preferred at the Manchester gas-works, where cannel coal only is employed.

and hold-fast screw, as shewn in fig. 8, and rendered air-tight by a luting, for which purpose the refuse lime from the purifier is found very convenient. Three, five, or seven retorts are arranged in one furnace or "bed," according to the extent of the works. Each retort has what is called a mouth-piece *a*, fig. 8, projecting from the front of the brickwork, from which rises the upright tube leading into the hydraulic main. The diameter of this tube is three or four inches, its height to the bend from ten to fourteen feet, and the distance of the hydraulic main from the bend from three to four feet. The bend is commonly made by a plain curve, but sometimes by a saddle-joint, as in fig. 8.

Fig. 8.

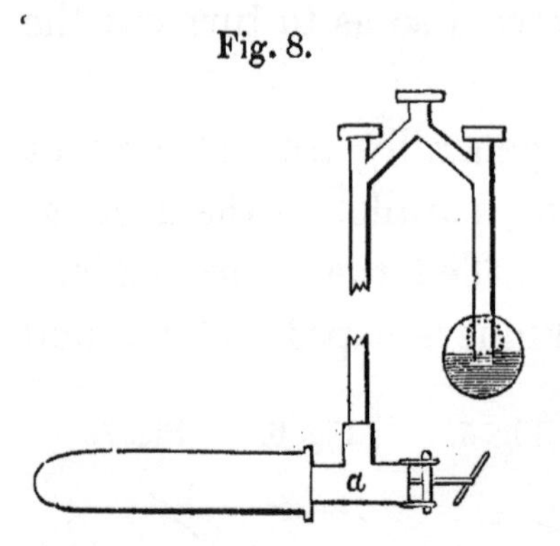

As the extremity of each tube from the retorts dips under the liquor in the hydraulic main, all direct communication between the gas in the latter and that in the retort is cut off, so that either of the retorts may be emptied and discharged while the others are being worked, without interfering with the process. The height of the exit-tube above the hydraulic main must be sufficient to prevent the liquor being driven into the retort by the pressure on the gas in the lime purifier *g*, fig. 1, page 64.

In extensive gas-works there are from four to six hundred retorts, of which from two to three hundred are worked on the average of summer and winter, each retort being charged with about a hundred and twenty pounds of coal every six hours. The retort furnaces are arranged in rows, generally on each side of the retort-house, the flues from the different furnaces meeting in a central chimney. As coke is the fuel generally employed to heat the retorts, little or no smoke is given off. In a well-conducted establishment, two men are sufficient for the management of three furnaces of five retorts each; but, as the retorts are kept in constant work throughout the twenty-four hours, relays of men are required for the night-work.

By the decomposition of the olefiant gas and vapours of hydrocarbons at a high temperature (page 57), the interior of the retorts becomes lined with a dense carbonaceous deposit, which, by its imperfect conducting power, offers an impediment to the transmission of heat. A patent was obtained in 1837, by Mr. Kirkham, engineer, for a mode of removing the incrustation by means of a jet of heated atmospheric air, which is impelled with force into the interior of the retort, maintained at a red heat during the operation. The air is conveyed by means of an iron pipe with flexible joints proceeding from a blowing machine, and bent so as to allow the nozzle, at its extremity, to be directed to any point on the surface of the interior of the retort.

The tube which conducts the tar and ammoniacal liquor to the cistern *d*, usually proceeds from the end of the hydraulic main, a little above the middle, so as to leave the latter always half full, or nearly so, of liquor. The tar and ammoniacal liquor are not miscible; and the former, being the heaviest, is found at the bottom of the cistern, from whence it may be withdrawn by the cock *i*.

Although a very large quantity of tar and ammoniacal liquor is deposited in the hydraulic main, yet the gas, in consequence of its high temperature, retains a considerable amount of these bodies in the state of vapour, which must be removed before it enters the lime purifier. To effect their separation, the temperature of the gas is reduced by passing it through the *cooler* or *condenser* (*f*, fig. 1).

Figs. 9 and 10 represent a convenient and effective condenser, which was invented and patented by Mr.

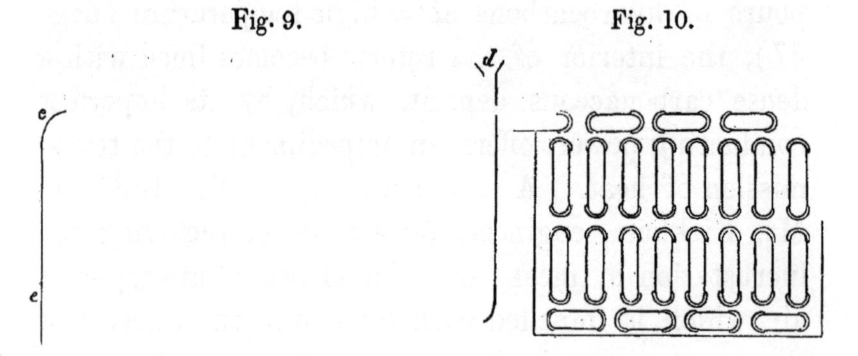

Fig. 9. Fig. 10.

John Perks, in 1817. It consists of a series of four-inch tubes, arranged in a rectangular iron chest, double-bottomed and open at top. The dimensions of the tank for a moderate-sized establishment may be ten feet in height, six feet six inches in length, and six feet in breadth. The bottoms of the tubes are open, and terminate in the plate *a*, between which and the true bottom are placed partitions, to separate the extremity *b* from the extremity *c*, *b*′ from *c*′, *b*″ from *c*″, &c.; but these partitions do not quite extend to the true bottom. The tubes are severally connected at top; either by a curved tube or a saddle-joint, so as to form, with the partitions at bottom, one continuous tube. The appearance of the top of the tank is shewn in fig. 10. A tank of the above dimensions may contain forty-eight upright tubes, each eight feet six inches in length, making, with the bends at top and communications at bottom, a continuous tube of a length of 432 feet. Cold water is poured into the cooler by the funnel-tube *d*, the heated water passing off by the pipe *e*. The tar and ammoniacal liquor fall into the space *f*, from which they are withdrawn by the tube *g*, and conducted, in some gas-works, into the tar-cistern. Sometimes, instead of enclosing the tubes in a cistern of water, they are made somewhat longer, and cooled by mere exposure to the air.

The condenser represented in fig. 1, page 64, is of a different construction from that just described, and not so generally adopted as the latter. This form, which was proposed by Mr. Malam, and is reported to be satisfactory in its operation, consists of an iron rectangular vessel about nine feet long, five feet wide, and four feet deep, containing several horizontal partitions or shelves of cast-iron plates, which have edges of about three inches in height, for the purpose of holding water. These shelves are all fastened to the sides and, alternately, to one end of the tank, a space of about six inches remaining between the other end of the shelf and the tank. Water is first poured in through an opening at top, and when the first shelf is full, the water flows over into the second shelf, and so on until all are filled. The excess of water flows out by the pipe *t*. The gas enters the box at *v*, and passing over successive sheets of water, equal to the area of each shelf, becomes thoroughly cooled, and passes out at *s*. The condensed matters flow out by the pipe *t*.

From the cooler the gas is next conducted to the lime purifier *g*, fig. 1, where it is freed from carbonic acid, sulphuretted hydrogen, hydrocyanic acid (or cyanogen), and some ammonia. (See table, page 65.)

The separation of the sulphuretted hydrogen is the most important object in all the purifying operations. This gas not only affords no light on combustion, but diffuses an intolerable odour if it escapes into the atmosphere unburned, is highly poisonous, and produces by its combustion sulphurous acid, which, by the action of the oxygen of the air and moisture, passes into the state of sulphuric acid, or oil of vitriol. The complete removal of sulphuretted hydrogen comes, therefore, to be a matter of no small importance. The only manner of effecting its separation is by the action of some substance for which it has a chemical affinity, but which is without action on the pure gas. A substance of this kind is *lime*, which was suggested as a purifier by Dr. Henry of Manchester in 1808, but was not used generally on the large scale until several years afterwards.

Lime not only absorbs sulphuretted hydrogen, but also carbonic acid and a little hydrocyanic acid, forming, respectively, sulphuret of calcium (hydrosulphate of lime), carbonate of lime, and cyanide of calcium. The principal methods employed to get rid of sulphuretted hydrogen, previous to the introduction of lime, were, passing the gas through hot iron pipes and through a large quantity of water; but such methods are quite ineffectual.

Lime may be applied to the gas in two ways: 1st, mixed up with water to about the consistence of thin cream (in the "wet-lime purifier"); and 2ndly, as slaked lime slightly moistened with water (in the "dry-lime purifier"). The former method is most commonly practised.

The wet-lime purifier represented at *g*, fig. 1, is a round cistern, into which the gas is conducted by the pipe *k*; this pipe is greatly expanded at its extremity into a cone and flat disc, and the latter is pierced with a great number of very small apertures: *l* is a rouser or agitator, to stir up the lime at the bottom of the purifier; it consists of an upright shaft, with a wooden or iron framework at bottom. The shaft of the rouser passes through a stuffing-box at the top of the purifier, and is turned by wheel and pinion work. In large gas-works the rouser is kept in motion by a steam-engine, which is also used for lifting coals, pumping up water, and other heavy work; but in small gas-works the rouser is worked by the hand. The milk of lime is made with one part of slaked lime to about twenty-five parts of water: it is introduced by the tube *p*, and withdrawn by the opening *r*, four times in twenty-four hours. The height of the liquid above the disc is usually about ten inches. In small gas-works only one lime purifier is used at a time; but in larger establishments the gas is passed through two, and sometimes three purifiers successively.

In some of the London gas-works the gas is passed through three or four purifiers at different elevations; the second stands at a higher level than the first, and the third higher than the second. By this arrangement the lime is made to enter and leave the purifiers in an uninterrupted current: the highest purifier, which is also the last through which the gas is passed, receives the milk of lime from a cistern placed above it, and, as the liquid attains a certain height, it is conducted by a discharge-tube, or waste-pipe, into the second purifier on a lower level, from whence the liquid is conducted into the first purifier. From the latter it is let off by an exit-pipe. As the gas travels in the opposite direction to the lime liquor, it is brought gradually into contact with purer lime: in the first vessel it meets liquor which has already purified other portions of gas in the vessels above; in the second vessel it meets a purer liquor; and, almost purified itself, passes from thence into the third vessel, where it meets lime fresh from the cistern above. A more effectual method of obtaining the complete absorption of the condensable gases could hardly be devised.

When the gas is arrived at the last purifier, to ascertain if all the sulphuretted hydrogen is absorbed, it is tested by carbonate or acetate of lead, which, like all other salts of lead, have the property of becoming brown or black when put into contact with sulphuretted hydrogen, the sulphuret of lead being formed. These tests are commonly applied by allowing a jet of the gas to escape, by opening a stop-cock fixed for the purpose to the cover of the purifier, and holding against the jet a card smeared with the carbonate or moistened with a solution of the acetate or subacetate of lead. If sulphuretted hydrogen remains, to the amount of no more than the thirty-thousandth part of the bulk of the gas, it may be easily detected by the coloration of the card. It is necessary that the card should be moist in this experiment, as some salts of lead thoroughly dry are not affected by sulphuretted hydrogen gas, if the latter is also perfectly dry. Another way of applying the lead test is, by attaching to the cover of the purifier one end of a bent tube, provided with a stop-cock, the other end of which is made to dip into a vessel containing an aqueous solution of acetate or subacetate of lead. If the gas is pure, it is commonly next conducted by the pipe *m*, to the gasometer *h*.

In the process of purifying the gas by slaked lime, merely moistened, in the "dry-lime purifier," patented by Mr. Reuben Phillips of Exeter, the gas is introduced at the bottom of a rectangular iron vessel (of the dimensions for five hundred lights, of five feet in length, five feet in breadth, and three feet in depth), and passed upwards through several layers of lime, placed on iron gratings or on perforated shelves of cast-iron, about seven or eight inches apart. The perforations are three-eighths of an inch in diameter, and three-quarters of an inch distant from centre to centre. The gas is conducted by a pipe from the top of the purifier to the gas-holder. Fresh slaked lime, moistened, but not sufficiently to be adherent to the hand, is placed on each shelf to the depth of three or four inches, and is then wetted with about a gallon of water from a watering-pot with a rose. The top of the tank is moveable, and fits into a water-joint, or trough, ten inches deep and six inches broad. The shelves are also moveable, so that the upper ones may be removed while the lower are being charged. This form of purifier is found very convenient in small works, but is too bulky when adapted to large establishments. Mr. Malam, however, patented, in 1822, an effective purifier for the use of extensive gas-works, on the same principle, but more complicated. This method is said to require nearly twice as much lime as the wet purifier, without the purity of the gas being increased; but it has a great advantage in requiring no constant mechanical power. In the wet purifier a bushel of lime is sufficient to purify, on the average, about twelve thousand cubic feet of gas.

The spoiled and fœtid lime was formerly allowed to run to waste, to the great annoyance of the neighbourhood; it is now preserved and employed as manure, and also as a cement to lute the covers of the retorts. The fœtid liquor which is drained from the lime obtained in the wet-lime purifier, is thrown into the ash-pits of the furnaces, where it evaporates, its vapour passing through the fire and up the chimney. By keeping the bars of the furnace cool, the vapour tends materially to their preservation.

The *gasometer*, *h*, fig. 1, consists essentially of two parts: a round cistern, open at top, and filled with water; and the gas-holder, which is a cylindrical vessel, open at bottom and closed at top, very little smaller in diameter than the cistern within which it floats.

The size of the gasometer varies from thirty to fifty feet in diameter, according to the extent of the works. It is considered that a capacity of thirty thousand cubic feet is the largest dimensions which

oxide; hence its feeble illuminating power. That produced at a bright red heat contains a larger proportion of olefiant gas and vapours of hydrocarbons than what is formed at any higher or lower temperature. As the distillation advances, the temperature being somewhat increased, the proportion of illuminating gases decreases considerably, while that of carbonic oxide and hydrogen increases in proportion. The density of the gas also, which is to a certain extent proportional to its illuminating power, gradually decreases with the duration of the process and the increase of the temperature. The following tabular view of results of the examination, by Dr. Henry, of the gas evolved from cannel coal at different periods of the process, will shew more clearly the difference in the composition.

COMPOSITION OF COAL-GAS IN 100 VOLUMES.

	1.	2.	3.	4.	5.
Olefiant gas and vapours of hydrocarbons	13	12	12	7	0
Light carburetted hydrogen	82·5	72	58	56	20
Carbonic oxide	3·2	1·9	12·3	11	10
Hydrogen gas	0	8·8	16·	21·3	60
Nitrogen gas	1·3	5·3	1·7	4·7	10
	100·0	100·0	100·0	100·0	100
Density of original gas (air as 1000) . .	650	620	630	500	345
Measures of oxygen required for combustion of 100 measures of gas . . .	217	194	196	166	78
Measures of carbonic acid produced . . .	128	106	108	93	30
Density of gas remaining after agitation with chlorine	575	527	533	450	345

Nos. 1, 2, and 3 of the preceding table were produced during the first hour of the distillation, No. 4 at the commencement of the sixth hour, and No. 5 ten hours after the commencement.

Some of the rules for the production of good coal-gas may be deduced from the preceding observations. In the first place, to reduce the product of tar as much as possible, the coal should be rapidly heated to bright redness; the retort, in fact, should always be at a bright red heat when the coal is introduced; a portion of the tar, which would otherwise distil over into the hydraulic main, is thereby converted into excellent gas. For the same reason, the coals should not be introduced into the retort in large masses. In that case, an exterior coating of coke is soon formed, which retards, by its non-conducting power, the transmission of heat to the interior of the mass. The advantage in the use of the D-shaped, kidney-shaped, and elliptical retorts, over the circular or cylindrical (page 67), is wholly referable to the circumstance, that the coal is more rapidly heated in the former than in the latter, from the exposure of a greater surface to the heat. The average quantity of gas obtained from a ton of Newcastle coal in the D-shaped, kidney-shaped, or elliptical retort, is 9000 cubic feet; in the circular retort, the diameter of which is about the same as the widest diameter of the other construction, the average quantity from the same kind of coal, at the same degree of heat, and worked in the same manner, is 6400 feet. According to the experiments of Mr. Peckston, the charge in elliptical, kidney-formed, or D-shaped retorts, may be worked off in half the time required with circular retorts. Five of the former, worked with 120 pounds of coal to each, during a four-hours' charge, will produce, in twenty-four hours, as much gas as ten cylindrical retorts, worked at eight-hour charges with 160 pounds of coal to each retort every charge. (PECKSTON, *Practical Treatise on Gas-lighting*, p. 126, 3d edit.)

The form of retort said to be best adapted for the rapid decomposition of the coal is the "*rotary retort*," patented by Mr. Clegg in 1816, but its expensiveness has hitherto proved an obstacle to its general introduction. It consists essentially of a large rectangular iron box, containing fifteen little boxes, each charged with about fifty pounds of coal. Only a portion of the large box is heated to redness, the temperature of the remainder being insufficient to cause the decomposition of coal. The small boxes remain for four hours in the cooler part of the large box, during which time the coal becomes thoroughly dried; they are then placed (by five at a time) for two hours at the part heated to redness. The gas obtained from the rotary retort is of excellent quality, and a ton of common coals is said to afford 11,400 cubic feet. Twenty-seven hundred-weight of coals may be worked off in twenty-four hours.

During the whole process, the temperature of the retort should not exceed that of a bright red heat. At a higher temperature, the proportion of hydrogen, carbonic oxide, and nitrogen gases is greatly increased; while the production of olefiant gas, the chief illuminating ingredient, almost ceases. This arises from the decomposition of the olefiant gas at a high temperature into free hydrogen or light carburetted hydrogen and carbon; the latter is deposited in a solid state around the sides of the retort in a peculiar form, known as gas-carbon. The carbonic oxide gas (CO or C_2O_2) is derived from the combination of carbonic acid gas (CO_2) with carbon; one equivalent of carbonic acid giving rise to two equivalents of carbonic oxide, by uniting with one of carbon. The nitrogen is probably derived immediately from the decomposition of ammonia into its elementary constituents, hydrogen and nitrogen. If both the carbonic acid and ammonia had remained as such, instead of being converted into carbonic oxide, nitrogen, and hydrogen, they might have been readily separated in the process of purification; but no efficacious and simple means are known of removing the above products of their decomposition. Hence the disadvantages of increasing the temperature beyond that of bright redness. As nitrogen and carbonic oxide gases are always disengaged in abundance at the end, in the ordinary manner of conducting the process, if gas of a very superior quality is required, the operation should be stopped two hours after the commencement.

But the quality and quantity of the gas depend, as might be supposed, on the quality of the coal, as well as on the manner of conducting the distillation.

The selection of a proper kind of coal for making gas is a subject which seems to deserve more attention from the managers of gas-works than it commonly receives. It is probable that in the end there would be a considerable saving, even in the London gas-works, by the adoption of richer kinds of coal than those commonly employed, as cannel coal, for instance, which would generate a more highly carburetted gas, and give as brilliant a light in smaller quantity, without the production of so much heat as is developed in the combustion of the highly hydrogenated gas derived from the commoner kinds of coal. In such case, the price of the gas per meter might reasonably be increased. Less labour and a smaller capital, too, are required to work a rich than a poor coal, as the apparatus for the former need not be on so extensive a scale as for the latter.

The following table shews the quantity of gas obtained on the large scale from several different kinds of coal in elliptical retorts at a bright cherry-red heat by daylight, the charge to each retort being about 126 pounds of coal.

Names of Coals.	*Cubic Feet of Gas obtained from One Ton of Coals.*
Scotch Cannel	11,850
Lancashire Cannel	11,680
Yorkshire Cannel	11,240
Bewicke and Craister's Wallsend	10,370
Russell's Wallsend	10,360
Bewicke's Wallsend	10,131
Tanfield Moor	10,070
Bell's Wallsend	9,963
Forest of Dean (High Delph)	9,880
Heaton Main	9,740
Hartley's	9,600
Cowper's High Main	9,460
Killingworth Main	9,393
Benton Main	9,082
Pontops	9,040
Wigan Ovall	9,000
Wear Wallsend	8,652
Burdon Main	8,341

Names of Coals.	*Cubic Feet of Gas obtained from One Ton of Coals.*
Brown's Wallsend	8,336
Wellington Main	8,270
Temple Main	8,180
Headsworth	8,052
Hebburn Seam	7,896
Hutton Seam	7,785
Nesham	7,763
Manor Wallsend	7,700
Forest of Dean (Low Delph)	7,660
Bleyth	7,420
Forest of Dean (Middle Delph)	7,260
Eden Main	6,670
Staffordshire Coal, first kind	6,474
Primrose Main	6,220
Staffordshire, second kind	6,090
Do. third kind	5,840
Do. fourth kind	5,807
Pembry	4,200

The results in the preceding table were the fruit of a series of careful and laborious experiments by Mr. Peckston, the greater part of which were performed on an extensive scale, and all sufficiently large for obtaining results of great practical value. (*Practical treatise on Gas-lighting.*)

In the process of purification by lime, to which the gas is subjected in order to deprive it of sulphuretted hydrogen and carbonic acid, the illuminating power of the gas appears to be reduced to a considerable extent; hence the advantage of selecting such kinds of coal as contain least iron pyrites. The reduction in illuminating power, where extensive purification is required, is shewn in the results of some experiments by Dr. Ure, who found, in a specimen of coal-gas, as delivered from the retorts of one of the metropolitan companies, no less than 18 per cent. of olefiant gas, or of the vapours of hydrocarbons and olefiant gas; but, after having been passed through the purifiers, there remained only 11 per cent. Mr. John Davies of Manchester estimates that 10 per cent. more of light might be realized by making use of a coal nearly free from sulphur, in the place of the common coal containing pyrites. (*Meeting of Brit. Association*, 1842.) There is reason to believe, however, that this estimate is too high for general application.

M. Penot has called the attention of the managers of gas-works to the great advantage which would result by the employment of coal previously deprived of its hygrometric moisture by drying. In its ordinary state, coal, according to M. Penot, contains 10 per cent. of hygrometric water. When steam comes into contact with olefiant gas and the vapours of hydrocarbons at a red heat, mutual decomposition takes place, with the production of carbonic acid (or carbonic oxide) and hydrogen or light carburetted hydrogen. The loss of the luminiferous constituents of the gas from this cause seems to be very considerable, but may be prevented to a great extent by making use of coal previously dried. It is stated that the quantity of olefiant gas produced from coal containing its ordinary proportion of hygrometric moisture, is to that produced from the same coal in a dry state as 1 : 1·5.

One kilogramme of coal, containing 10 per cent. of water, afforded

160 litres of gas of good quality, giving a white flame,

92 litres of inferior gas, giving a red flame,

———

252

leaving 632 kilogrammes of coke.

One kilogramme of coal, previously well dried, afforded
240 litres of good gas,
92 litres of inferior gas,

332
leaving 668 kilogrammes of coke.

The difference is certainly far greater than might have been anticipated, but the exactitude of the proportions determined by M. Penot has been confirmed by experiments made on an extensive scale at the Mulhausen gas-works by a committee appointed by the *Société Industrielle* of Mulhausen. M. Penot's paper on this subject may be found in the *Journal für praktischen Chemie*, xxiv. 106, and the report of the committee in the *Bulletin* of the society.

The presence of *ammonia* in coal-gas, after having been subjected to the ordinary processes of purification, seems to have escaped notice until lately. A considerable quantity of the sulphuretted hydrogen which enters the lime purifier is in a state of combination with ammonia, as hydrosulphate. This substance is decomposed by the lime in the purifier, with formation of sulphuret of calcium and evolution of free ammonia. Carbonate of ammonia also enters the lime purifier in the state of vapour, and is decomposed by the lime in a similar manner, with production of carbonate of lime and free ammonia. A small proportion of the ammonia formed by these sources remains dissolved in the water of the purifier; but the greater part comes off as gas, mixed with coal-gas. This impurity is not very injurious to the illuminating power of the coal-gas, and, if the gas is kept in the gasometer for a short time, the water abstracts it almost entirely; but several processes have been proposed for the removal of the ammonia before the gas enters the gasometer. The process patented by Mr. Phillips of Exeter consists in conducting the gas from the lime purifier into a tank containing a solution of alum. In order to expose a large surface of the purifying liquid to the gas, a quantity of broom is immersed in the solution. In the course of a few hours the bottom of the purifier becomes covered with a quantity of alumina, produced through the decomposition of the alum. The sulphate of alumina contained in the alum produces, with free ammonia, sulphate of ammonia, which remains in solution, and alumina, which is precipitated.

A solution of protosulphate of iron (green vitriol) was substituted for alum in the above process with a corresponding result, oxide of iron being precipitated, and sulphate of ammonia formed in the solution; but green vitriol presents advantages which are not possessed by alum. Hydrosulphate of ammonia and hydrocyanic acid, traces of which sometimes escape the action of the lime in the purifier, are imperfectly absorbed by a solution of alum: the latter absorbs only the ammonia of the hydrosulphate; the sulphuretted hydrogen passes on to the gasometer. But green vitriol absorbs hydrosulphate of ammonia as well as free ammonia, producing in the former case sulphuret of iron and sulphate of ammonia, instead of oxide of iron and sulphate of ammonia, as happens with free ammonia only. The hydrocyanic acid, or, more properly, hydrocyanate of ammonia, is likewise absorbed by green vitriol, with the production of a variety of Prussian blue.

Mr. Lowe, of the Chartered Gas Company's works, proposed to separate the ammonia by passing the gas through dilute sulphuric acid, after having passed the lime purifier. As concentrated sulphuric acid (oil of vitriol) rapidly absorbs the vapours of the oily hydrocarbons, and also olefiant gas, it cannot be used for this purpose; but, when the dilute acid has been so long in use as to be saturated with ammonia, a small quantity of oil of vitriol may be introduced by a funnel tube, fitted to the purifying vessel for that purpose. When it is required to separate the ammonia on the small scale, as by the consumer himself, the gas may be exposed to the dilute acid in Mr. Lowe's naphthalising box with perforated shelves, a description of which may be found in another part of the present article.

Another method of separating the ammonia was proposed by M. Blondeau de Carolles, which consists in conducting the gas from the lime purifier through a vessel containing layers of coke covered with chloride of calcium, a substance which rapidly absorbs ammonia.

But neither of these purifying agents is much employed at present, as it is found that the ammonia may be very easily and completely separated by merely washing the gas with water. The most effectual method of washing the gas is to conduct it from the purifier to the bottom of a tank containing broom-twigs, water being introduced at the top and allowed to trickle slowly over the broom. The ordinary wet-lime purifier (*g*, fig. 1) may also be used for the same purpose. The solution of ammonia which is thus obtained is employed in the manufacture of sal-ammoniac and sulphate of ammonia.

Mr. Mallet has proposed to transmit the gas, as it comes from the condenser, through two purifiers, containing solutions of green vitriol or solutions of sulphate of manganese,* and afterwards through the milk of lime. The sulphate of ammonia thus obtained would not only cover the expense of the green vitriol, or sulphate of manganese, but give a clear profit. It is said that very little sulphuretted hydrogen remains to be absorbed by the milk of lime, and that the gas thus purified does not produce a trace of sulphurous acid in its combustion. An incidental advantage attending this method is, that much less labour is required to keep the lime in a state of agitation. In 1841 a patent was granted to Mr. Croll, superintendent of the Brick Lane gas-works, for the use of a solution of chloride of manganese, dilute sulphuric acid, and dilute muriatic acid, for a similar purpose: the gas is afterwards passed through the lime purifier.

Sulphate of lead, which is a bye-product in calico print-works, has been proposed by M. Penot as a purifying agent in districts where it may be procured at a very cheap rate. It absorbs hydrosulphate of ammonia completely, the oxide of lead retaining sulphuretted hydrogen, and the sulphuric acid am-

* Sulphate of manganese is obtained as a bye-product in the manufacture of bleaching-powder.

monia. In no part of this country, however, can sulphate of lead be procured sufficiently cheap for such an application.

§ V. SECONDARY PRODUCTS OF THE COAL-GAS MANUFACTURE.

Next to the gas itself, the most important and valuable of the volatile products of the distillation of coal is the ammoniacal liquor or gas-water. This liquid and the tar are condensed, for the most part, in the hydraulic main *b*, fig. 1, from whence they are conducted into the tar-cistern. A further quantity is also separated from the gas in the condenser. The tar is found at the bottom of the cistern, the ammoniacal liquor floating over it.

The ammoniacal liquor of the coal-gas works is at present the principal source from whence the commercial demand for sal-ammoniac and carbonate of ammonia is supplied. The average price of this liquor may be taken at half-a-crown per butt of 108 gallons. A ton of good coals affords rather more than two hundred pounds of ammoniacal liquor.

This liquid is essentially an aqueous solution of hydrosulphate and carbonate of ammonia; but it also contains muriate of ammonia, sometimes in considerable quantity,* hydrocyanate of ammonia, acetate of ammonia,† gallate of ammonia (Mr. Leigh), and sul-

* Four ounces of muriate of ammonia have been obtained from one gallon of gas-water from cannel coal (Mr. Leigh).

† Acetate of ammonia is a constant product, according to Lampadius.

phite of ammonia. Muriate of ammonia is prepared from this source by saturation with muriatic acid, when sulphuretted hydrogen, carbonic acid, and a little hydrocyanic acid are given off, and an equivalent quantity of muriate of ammonia remains in solution. By evaporation to dryness and sublimation this salt is obtained nearly pure. Sulphate of ammonia is prepared in very considerable quantity from ammoniacal liquor; not to be used itself, but to be afterwards converted into carbonate of ammonia (smelling-salts). The sulphate is made either by saturating the liquor with sulphuric acid, or by the addition of green vitriol, procured by the oxidation of iron pyrites. In the latter case, sulphuret and carbonate of iron are first formed and precipitated, being insoluble compounds. Sulphate of ammonia remains in solution, and is obtained by evaporation and crystallization. At the Deptford chemical works, which is one of the principal manufactories of ammoniacal salts from gas-liquor in this country, there is a yearly consumption of between five and six hundred thousand gallons of the liquor. A pipe is laid from the Deptford gas-works to the above establishment for the purpose of conveying the liquor.

Of late years it has been proposed to make the hydrocyanate of ammonia contained in gas-liquor available as a source of Prussian blue. The common method of preparing that pigment from gas-liquor is first to saturate the liquor with muriatic acid, and to add afterwards a solution of green vitriol: but this method presents no economical advantages over the ordinary process.

The proportion of tar obtained from coal in the process of making gas generally amounts to about 8 per cent. of the coal. The disagreeable smell of this substance is an impediment to its extensive employment; but it is used as a paint, to protect wood, &c. from injury by moisture. When distilled, one hundred pounds of tar afford about twenty-six pounds of an oily liquid known as *coal-oil;* the light product which first distils over is *coal-naphtha.* A black resinous substance, *pitch*, remains behind, of the chemical nature of which little is known: it is largely used for paying wooden piles, &c. which are to be immersed in water, and the bottoms of ships; but it is not so well adapted for these purposes as pitch obtained from wood-tar; it may be procured, however, at a cheaper rate than the latter.

Coal-naphtha is a mixture of several bodies, some of which are neutral, some possess the properties of an acid, and others those of a base. The most remarkable constituent is naphthalin, the properties of which, together with those of paranaphthalin, have already been described. Under the names of carbolic acid, rosolic acid, brunolic acid, pyrrol, leucol, and cyanol, M. Runge has described some of the compounds said to be contained in coal-naphtha, but some of these bodies were probably products of the decomposition of the true ingredients. According to M. Dumas, benzin forms an ingredient of this complicated mixture, and M. Laurent believes a substance which he has named the hydrate of phenyle to be an essential constituent.

Coal-naphtha is a substance of considerable importance in the arts, it being one of the best and most available solvents we possess for caoutchouc. It is also burned for the production of light in a peculiar kind of lamp, known as the naphtha-lamp, and is employed to impregnate coal-gas with its vapour, by which a considerable increase in the illuminating power of the gas is obtained. (The means of effecting the impregnation of coal-gas with the vapour of naphtha will be afterwards described.)

Tar may be viewed as an intermediate product in the conversion of coal into gas. It is given off from the coal at a comparatively lower tempera-ture than gas, and is always the first product of the decomposition of coal. To reduce the product of tar, it was proposed by Mr. Parker of Liverpool to pass the gas, as it issues from the coal retorts, through iron tubes heated to bright redness, by which the quantity of gas would be greatly increased and that of tar diminished; but the olefiant gas would become converted to light carburetted hydro-gen, with deposition of carbon, by long exposure to a bright red heat. The illuminating power of the gas would thereby be diminished, although the volume of the gas would be increased.

In some gas-works tar is made available as fuel for heating the retorts. About forty gallons are found to be sufficient for a ton of cannel coal.

The coke which remains behind in the retort in the process of gas-making is not the least valuable of the secondary products. It consists of carbon to the amount of from 75 to 95 per cent., the remainder being the fixed saline and earthy matters of the coal. The proportion of coke formed generally amounts to about two bushels from a hundred-weight of coal, or 60 per cent. on the original weight of the coal. Coke is always considerably lighter than the coal from which it is produced; the bulk of the coke formed from those kinds of coal which are used for making gas in this country is generally 30 per cent. greater than that of the coal. In its physical aspect coke presents as many differences as exist in the varieties of coal from which it is produced. Some kinds of coal produce an infusible, and others a fusible coke, or rather a coke which fuses before it becomes fully carbonized. The coke from the former kinds of coal has the same shape as the original masses; but the coals which produce a fusible coke swell up considerably before becoming fully carbonized, the coke retaining the expanded form. Coals which produce an infusible coke give, on destructive distillation, more water and less tar than the fusible coals.

Coke is a reducing agent of great importance in certain metallurgic operations, as it produces by its combustion a higher temperature than any other fuel, bulk for bulk. It is estimated in metallurgic works that one part of coke by volume will afford as high a temperature as two parts by volume of wood charcoal; by weight, one part of the latter is considered equal to from one and a quarter to one and a half parts of the former. To produce a given effect, there is always consumed more of coke than of wood charcoal by weight, especially if a very elevated temperature is not required. The reason of this is that coke does not burn well, except in large masses and with a powerful current of air, by which some of the heat is quickly dissipated. The earthy impurities contained in coke are a serious inconvenience in its employment in certain operations in metallurgy and the assaying of ores.

§ VI. OIL-GAS.

In a few localities, where coal is difficult to procure, oil may be advantageously substituted as a source of gas. The oil employed for this purpose is the crudest and cheapest that can be procured; even pilchard dregs and the sediment of whale-oil, quite unfit for burning in the ordinary manner, are sufficiently pure for making gas. But oil is not at present used in the manufacture of gas on the large scale in any place where coal is accessible at a moderate price.

The process of making oil-gas is much more simple than that of coal-gas, as the former requires far less purification. In one circumstance the two processes differ essentially. The oil is not introduced in quantity into the retort and heated for several hours, as coal is; in such a case the greater part of the oil would distil over without undergoing much alteration, and that portion only would be converted into combustible gas which is in immediate contact with the heated sides of the retort. What is required is the means of bringing a small quantity of the oil rapidly to a high temperature, even a red heat; this is effectually attained by means of a simple apparatus invented by Messrs. J. and P. Taylor in 1815.

The original apparatus consisted merely of a furnace with a contorted iron tube, containing fragments of brick or coke, into which, when red-hot, the oil was allowed to drop.

When heated in this manner, oil is immediately decomposed, gases are given off, accompanied with a considerable quantity of vapours of substances which are liquid at common temperatures, and a large deposit of carbon takes place in the retort tube.

An improvement has been since effected in the apparatus, which consists in conducting the exit-tube from the retort into an air-tight cistern or receiver, in which the more easily condensed products are collected in the liquid state and returned to the retort. The gas proceeds from a pipe leading from the top of the cistern. In the annexed figure, *a* represents the retort, filled with pieces of coke about the size of a hen's egg; *b* is the exit-tube, leading into the cistern *c*, from which the pipe *d* proceeds to the gasometer. The oil, melted fat, or distilled liquid of the cistern is introduced into the retort by the tube *e*. The coke is changed every fortnight or three weeks, as the interstices become obstructed by the deposit of carbon.

Fig. 12.

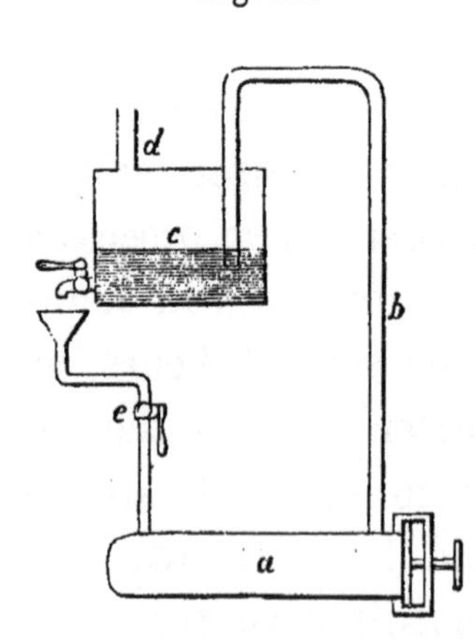

Oil-gas contains neither nitrogen nor sulphuretted hydrogen. It contains more carbonic oxide than coal-gas; but the presence of that gas (which has hardly any illuminating power) is more than counterbalanced by the large relative proportion of olefiant gas, on which the illuminating power of both oil-gas and coal-gas essentially depends. From the presence of so much olefiant gas and vapours of hydrocarbons which are liquid at common temperatures, the illuminating power of oil-gas is reckoned at three times that of ordinary coal-gas, and twice that of the best coal-gas. The specific gravity of good oil-gas is 900 ; but it varies much, according to the temperature at which the gas is produced. The proper heat is that of dull redness ; if lower, the gas contains too large a proportion of the vapours, which, instead of being condensed in the cistern, are condensed gradually in the pipes and gasometer, occasioning not only a great loss of product, but no small inconvenience from the stoppage of the pipes. On the other hand, if the heat is above dull redness, the bulk of the gas is greater; but its density and illuminating power are less, the olefiant gas being decomposed into carbon and light carburetted hydrogen, as before explained. The following table by Dr. Henry exhibits the different qualities of gas produced from oil at different temperatures.

COMPOSITION OF OIL-GAS IN 100 VOLUMES.

	No. 1.	*No.* 2.	*No.* 3.	*No.* 4.
Olefiant gas and vapours	38	22·5	19	6
Light carburetted hydrogen	46·5	50·3	32·4	28·2
Carbonic oxide gas	9·5	15·5	12·2	14·1
Hydrogen gas	3	7·7	32·4	45·1
Nitrogen gas	3	4	4	6·6
Oil-gas	100	100	100	100
Density	906	758	590	464
Volumes of oxygen required for combustion of 100 volumes	260	220	178	116
Volumes of carbonic acid produced	158	130	100	61

No. 1, was made from train-oil in the ordinary manufacturing process at a dull red heat ; No. 2, at a higher temperature ; and Nos. 3 and 4, higher still.

One gallon of whale-oil affords on the average about ninety cubic feet of gas of the specific gravity 900. A gallon of palm-oil afforded ninety-five feet.

When oil-gas is subjected to a pressure of from twenty to thirty atmospheres, as in the process formerly conducted on an extensive scale for rendering gas portable, about one-fifth of the bulk of the gas is condensed into an oily, volatile liquid, having a specific gravity of 821 ; when the pressure is removed, this liquid does not entirely re-assume the vaporous state, and it may be preserved in ordinary well stoppered bottles. Mr. Faraday ascertained the condensed liquid to be a mixture of several hydrocarbons differing considerably in their degrees of volatility. The boiling point of one is under the freezing point of water, and is therefore gaseous at common temperatures. This body has the same composition as olefiant gas in 100 parts, but twice the density. The principal constituent is a substance now known by the name of benzole, or benzin, which was found by Mr. Faraday to consist of carbon and hydrogen in the proportion of two equivalents of the former to one equivalent of the latter, and was hence called bicarburet of hydrogen. It was afterwards observed by M. Mitscherlich to be the principal product of the decomposition of crystallized benzoic acid when heated to redness in contact with lime. It is a limpid colourless liquid having the specific gravity 850. It freezes at 32° into a white crystalline mass, which melts again at 44° Fahr. Its boiling point is 186°; and its specific gravity, in the form of vapour, at 60° is 2738. Its odour is peculiar, but not disagreeable. It is soluble in ether and alcohol, but insoluble in water. It burns with a bright flame and much smoke; and, like most other of the denser hydrocarbons, when passed through a red-hot tube is resolved into carbon and light carburetted hydrogen. Its probable formula is $C_{12} H_6$.

Notwithstanding the great illuminating power of oil-gas, the abundance of the product compared with that from coal, and the simplicity of the process, yet the commonest oil is far too expensive in this country to rival coal as a source of gas. Several extensive oil-gas establishments were erected at different places in Great Britain, but all these have gradually become converted into coal-gas manufactories. Not a slight objection to oil-gas is the gradual liquefaction which its most valuable constituents undergo on standing with exposure to moderate cold, to the great deterioration of the gas and the clogging of the pipes.

§ VII. RESIN-GAS.

Resin is another substance which may be advantageously substituted for coal in the manufacture of light-gas where coal is not readily accessible. It affords an abundance of gas of excellent quality, nearly equal to oil-gas; but the price of resin compared with that of pit-coal must ever prevent the former from coming into successful competition with the latter in Great Britain. The attempt made, a few years ago, to introduce the manufacture on the large scale into this country proved a decided failure; but in some parts of France resin-gas is, I believe, manufactured with success.

The apparatus for making resin-gas is much the same as that for oil-gas. In the first attempts to obtain gas from resin the oil-gas apparatus was used without any modification; melted resin being caused to trickle as oil on the fragments of coke. But the exit-tube of the retort, which was carried to a considerable height to allow the return of some of the condensed volatile oil, became clogged by the bituminous matter which distilled from the resin. Professor Daniell overcame this difficulty by conducting the exit-pipe from *under* the retort into a cistern or hydraulic main, by which the return of the bituminous matter is effectually prevented. The construction of the apparatus as erected at Bow, near London, under the superintendence of Professor Daniell, for the late Resin Gas Company, will be understood with the assistance of the annexed figure. The retort *a* is charged with fragments of coke, and heated to bright redness by the furnace underneath. The exit-tube *b* passes into an air-tight cistern or hydraulic main *c*, which is kept cool by a

Fig. 13.

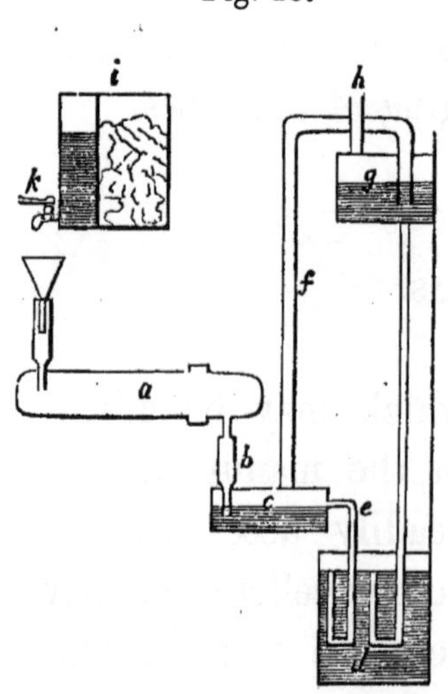

refrigerator, and supplied with water from a cistern above. The volatile oil, which condenses in *c*, is conducted into another cistern *d*, by the discharge-pipe *e*. The uncondensed gases and vapours pass up the pipe *f*, and deposit more volatile oil in the cistern *g*, from which it is conducted into *d* by a syphon-tube. From the cistern *g* the gas passes by the pipe *h* into the gasometer or other reservoir.

The large quantity of volatile oil which is collected in this process is employed to dissolve the resin, instead of melting the latter alone. The commonest resin of commerce is put into the vessel *i*, and mixed with the oil in the proportion of about eight or ten pounds of resin to a gallon of oil ; the solution of the resin is assisted by the heat of the furnace below. The partition shewn in the vessel *i* is a wire-gauze screen, to prevent any solid resin or impurity from entering the retort. The dissolved resin passes through the stopcock *k*, funnel, and syphon-tube, into the retort.

The distilled oil contains a little acetic acid, formed during the process, which it is necessary to neutralize by the addition of lime before the oil is mixed with the resin. The quantity of oil produced is a little more than what is required for the solution of the resin.

It is stated that a hundred-weight of resin affords, in a properly conducted operation, from one thousand to twelve hundred cubic feet of gas of the average specific gravity 850. The illuminating power of resin-gas compared with coal-gas is as two to five ; that is, two cubic feet of resin-gas afford as much light as five cubic feet of coal-gas. At an establishment in France the illuminating power of resin-gas compared with coal-gas is estimated as five to nine, one kilogramme of resin producing 497 litres of gas. The distilled oil is almost wholly transformed into gas if quickly heated to redness.

Turf-gas.—A series of experiments was conducted at the gas-works at Newry (Ireland), by Mr. Peckston, in 1823, in order to ascertain whether gas could be obtained from turf; from which he was led to conclude, that, where that substance is abundant, it may be advantageously substituted for coals in the production of light-gas. According to M. Merle, the light of the gas obtained from turf is more brilliant than that of coal-gas (?). The gas is easily purified, and the charcoal which remains in the retorts answers perfectly for domestic purposes; it gives considerable heat, and inflames readily.

§ VIII. MODE OF BURNING GAS.

The quantity of light obtained by the combustion of gas is greatly influenced by the mode in which it is burned. The truth of this statement will appear from the consideration of the structure of an ordinary flame.

A solid substance which does not afford a gas when heated strongly—a piece of iron or of charcoal, for instance,—presents at a red heat the phenomenon of *ignition*, but it never produces flame. All combustible bodies, on the contrary, which are either gaseous themselves, or are convertible, partially or entirely, into a gas, as hydrogen, sulphur, and phosphorus, always burn with flame. Hence flame might at first be regarded as *luminous gaseous matter*, but we shall immediately see that such a definition can only be taken in a very limited sense.

It is doubtful whether pure gaseous matter is, under any circumstances, capable of becoming luminous. If a gas is capable of emitting light when heated, it is certainly no more than a sensible glow, although the temperature may be sufficiently high to heat a metallic wire to whiteness. The air which rises from a gas-flame at an Argand jet is hot enough to heat a fine metallic wire to bright redness an inch or two above the highest part of the flame. The flame of pure hydrogen gas is so feebly luminous as to be scarcely visible in broad daylight; yet its temperature is sufficiently high to heat to whiteness a piece of thin platinum wire, or particles of lime thrown into the flame. The luminosity of flame, in fact, does not depend altogether on its temperature; but on the presence of solid matter diffused through the flame, and ignited by it, the solid particles acting as radiating points. Hence a luminous flame has been justly described by Davy as always containing *solid matter heated to whiteness.* No flame possesses less light than that of the oxyhydrogen blow-pipe; but, on introducing into it some solid infusible substance, the light that is emitted is too intense to be borne by the naked eye. The feeble luminosity of the flames of hydrogen, sulphur, and carbonic oxide is owing to the products of the combustion,—namely, water, sulphurous acid, and carbonic acid,—being gaseous at the temperature of the flame. On the other hand, the flames of phosphorus and of metallic zinc are very intense, because the products of the combustion—namely, phosphoric acid in the case of the former, and oxide of zinc in the case of the latter,—are solid, and being diffused through the flame serve as radiating points for the light and heat. If the luminosity of flame depends on the presence of solid matter, where then is the source of the solid matter in the flame of gas or an ordinary combustible ?

The combustion of coal-gas—that is, the conversion of its hydrogen into water, and its carbon into carbonic acid, through the combination of these elements with the oxygen of the air,—can of course only take place where the gas or vapour is in contact with the air. In an ordinary flame, such as that of a spirit-lamp, tallow-candle, or gas issuing from a plain jet, the gaseous matter is in free contact with the air only at the exterior of the flame, where alone perfect combustion takes place. In the interior, combustible gases still unburned exist. That such is the structure of the flame may be readily seen by depressing upon it a sheet of wire-gauze of small mesh (through which the flame is unable to pass), so as to view the flame in section. The combustion is then seen to be limited to the margin of the flame. But if the structure of the flame is examined a little more minutely, it is seen to consist of four parts, three of which are represented in the annexed figure. 1°, The shaded interior represents the unburned combustible gas ; 2°, around this is the brilliant part of the flame, or the flame,

Fig. 14.

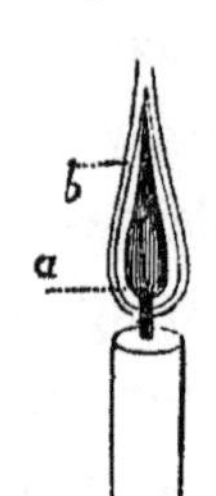

strictly so called; 3°, on the very exterior is another portion, hardly visible, but which may be perceived with attention in the flame of a tallow-candle; and 4°, the blue flame at the bottom constitutes another distinct portion.

In the blue part at the bottom, and on the very exterior of the flame, perfect combustion of the gas takes place, the whole of the carbon and hydrogen uniting with the oxygen of the air to form water and carbonic acid, and without the production, therefore, of solid matter to afford light. But in the luminous part of the flame, *b*, the supply of air is insufficient for complete combustion. The carburetted hydrogen gas in this part is decomposed by the heat into free carbon and hydrogen, as when passed through a red-hot porcelain tube; the hydrogen alone, or principally, burns, while the carbon is deposited in minute particles, which become heated to whiteness. The carbon does not burn entirely until it reaches the exterior of the flame. Thus the light of the flame of ordinary combustibles depends on the consecutive combustion of the hydrogen and carbon. The combustion of the hydrogen and some of the carbon serves to produce the requisite heat, and the particles of the remaining carbon radiate the light. The hottest part of the flame is just at the top of the luminous cone, where the combustion is perfect, but the air not in sufficient excess to carry away the heat quickly, as is the case in the exterior of the flame at the side and at the blue part at bottom. The heat in the centre of the flame is so low that gunpowder may be held there without being ignited; even fulminating silver has been held for some seconds in the interior of a flame without explosion. The chemical effects produced on many substances by the exterior and interior parts of the flame are exactly opposite. Just within the apex of the luminous cone certain metallic oxides are rapidly deprived of their oxygen and reduced to the metallic state ; but at the very summit of the flame, or a little beyond, the metals reduced from the same oxides re-absorb oxygen, reverting to the state of oxide. If a piece of clean copper wire is held at the very summit of the flame, it rapidly becomes covered with a coating of oxide ; but, on depressing the wire into the interior of the flame, the oxide is reduced, and the wire again becomes bright.

That the luminosity of the flame of ordinary combustibles depends on the deposition of solid carbon, through the consecutive combustion of the carbon and hydrogen, may be proved by some interesting experiments with coal-gas. If that gas is mixed with an equal volume of air before being burned, it deposits a smaller proportion of solid carbon, and loses half of its illuminating power. On increasing the quantity of air the light is still further diminished. If a sheet of fine wire-gauze is held close to the orifice of a gas-burner, and the gas kindled above it, it will of course burn with its ordinary brilliancy : but if the gauze is gradually elevated, so as to mix the gas with air in different proportions before it passes the gauze and burns, it will be found that the brilliancy of the flame diminishes with the height of the gauze ; that is, with the quantity of air mixed with the gas. At a particular elevation the gas burns with a faint blue flame, not more luminous than that of sulphur, and hardly visible in the direct rays of the sun. It is then mixed with the quantity of air necessary for the complete combustion both of the carbon and hydrogen simultaneously. The heat of such a flame is more intense than from the same quantity of gas burned in the ordinary manner, because the combustion takes place within a smaller compass in the former than in the latter. The intensity of the heat is seen by projecting into it solid particles of some fixed matter, as oxide of zinc, which instantly become heated to whiteness.

The conditions requisite to obtain the full amount of light from a flame may be deduced from the consideration of the preceding facts. In the first place, the carbon and hydrogen must be burned completely, though not at the same place. When the combustion of the carbon is incomplete, the flame is smoky, and a large proportion of light is lost. As this arises from the gas being in too large a proportion compared with the air, the remedy is obviously either the reduction of the amount of gas, or the increase of that of the air.

The great object of almost all improvements in the form of gas-burners and chimneys is to afford the means of increasing and regulating the amount of air brought into contact with the *surface* of the flame, so that the *intensity* of the combustion may be increased, without any derangement of the *order* of the combustion. From a plain jet, the diameter of the orifice of which is a quarter of an inch, a flame higher than two inches and a half cannot be obtained without smoke; but the same amount of gas which would give a smoky flame from a plain jet may be made to burn with a clear and brilliant flame by extending or dividing the aperture of the jet, so as to give the flame a larger surface in contact with the air. The entire superficies of the aperture need not be enlarged; it may, on the contrary, be diminished. In the place of a single aperture, the gas may issue, for example, from two holes drilled obliquely to make a cross flame, or from three holes, as fig. 15, which gives a flame called the "cockspur." More light is obtained by such a jet than by the single aperture; but the intermixture of air with the gas at the lateral jets prevents the deposition there of the proper proportion of carbon, so that the full amount of light is not obtained. It is to be observed that the supply of air is required merely at the surface of the flame, and in no larger quantity there than is requisite to maintain a proper combustion.

Fig. 15.

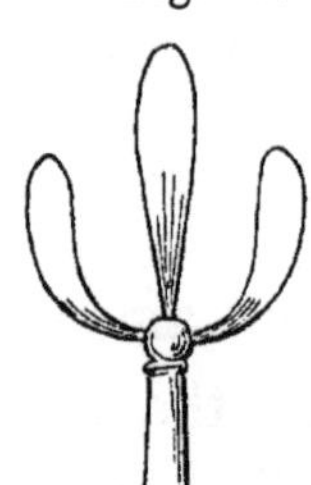

When the same burner has several apertures, the light is improved by making them so near to each other that all the jets of gas shall unite laterally into one, as in fig. 16, which represents a semi-circular jet of several holes, the flame from which is called the "union jet" or "fan;" or as fig. 17, the aperture for which is a slit across the top of the beak; the latter gives a sheet of flame called the "bat-wing." Of burners of this kind, that which seems to afford most

Fig. 16.

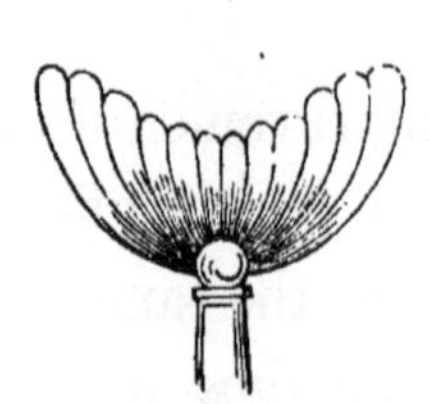

light by the consumption of the same quantity of gas, is the "fish-tail" or "swallow-tail" burner. In this the gas issues from a single small round hole in the centre of the top of the burner, but the single jet which is emitted is formed by the union of two oblique jets within the burner immediately before the gas is emitted. The result is a perpendicular sheet of flame, much the same in appearance as the "bat-wing" (fig. 17), but presenting somewhat more light at the bottom, on account of the smaller intermixture of common air with the gas at that part of the flame. The "fish-tail" burner and "bat-wing" are much more extensively used than the "fan," "cockspur," or any other of this description.

Fig. 17.

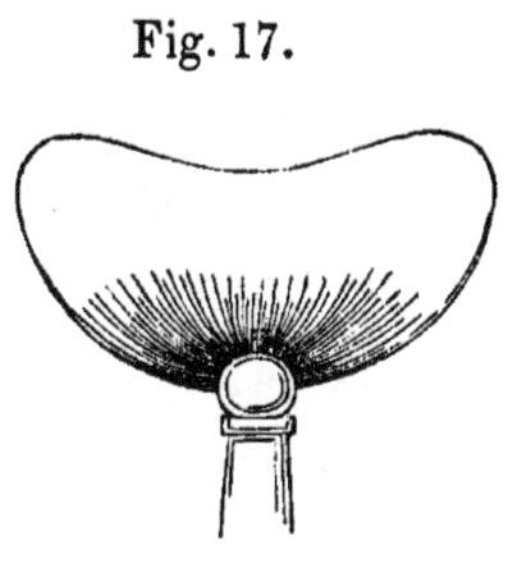

Another method of increasing the proportion of air is by the burner known as the Argand, which allows the passage of a current of air through the centre of the flame, as in the Argand oil-lamp. This burner consists of a circle of small holes of equal size, the centre of the circle being open to admit an upward current of air. The view of this burner, from above, is shewn in fig. 18, and in perpendicular section in fig. 19. The internal current of air thus supplied to the flame causes more combustion in the same time than with the plain jet, and therefore a higher temperature, but a sufficient quantity of solid carbon is still deposited to radiate the light. The same amount of gas, which,

Fig. 18. Fig. 19.

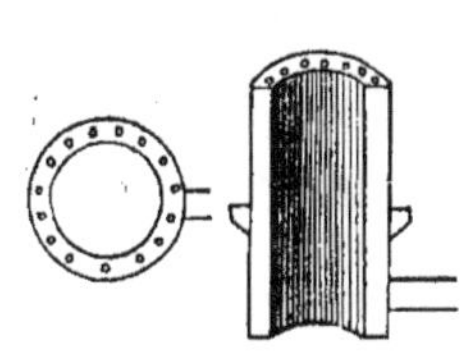

proceeding from a plain jet, would give a very tall, smoky, and dull flame, would give a much shorter and more luminous flame from an Argand burner, without any production of smoke. The increase of light is far greater, if, in the place of air, pure oxygen gas is passed up the internal opening, as in the original* Bude light of Mr. Goldsworthy Gurney. The intensity of the combustion on the surface of the flame is greatly increased by the oxygen, but the consecutive order of the combustion is by no means disturbed.

For the same consumption of gas, the flame from an Argand burner of ordinary construction does not afford so much light as the flat flames from the "fish-tail" and "bat-wing" burners. The reason of this is obvious: as luminous flame is opaque, each side of the interior surface of an Argand flame diffuses light into an apartment only above or below the flame on the opposite side. If the diameter of the circle of the burner does not exceed an inch (which may be taken as the average size), only a trifling amount of light is obtained from the interior surface of the flame.

When a great deal of light is required, the most economical burner seems to be a large circle of "fish-tail" burners, with as little metal-work at bottom as possible, in order to allow the radiation of light from the interior surfaces underneath the opposite sides. The light from such a burner may be greatly increased by adjusting to the top of the flame a horizontal

* The light now known as the Bude light is nothing more than an ordinary gas-flame from two, three, or more large concentric Argand burners, with chimneys and reflecting apparatus of particular construction.

reflector formed of a circular piece of metal with a round opening in the centre, over which a short glass chimney should be placed. The lateral union of the "fish-tail" jets should be avoided, as smoke is thereby produced; and if a profuse light is required, other jets may be placed behind the intervening spaces, so as to form a double circle. A burner of this construction was first introduced, I believe, by the Liverpool New Gas and Coke Company, and named the "solar gas-lamp."

When a considerable quantity of gas is burned together in one flame without smoke, the increase of light is much greater than the increase in the amount of gas consumed; hence, when a great deal of light is required, there is always a saving in burning a considerable quantity of gas together. An Argand burner supplied with

Cubic feet of gas per hour,		
$1\frac{1}{2}$	gave light equal to	1 candle,
2	"	4 candles,
3	"	10 candles.

To obtain the full amount of light from any burner, the flame should always be made as large as possible without smoking.

The construction of the chimney may have a marked influence on the proportion of air brought into contact with the surface of the flame. A very wide, short, and cylindrical chimney hardly increases or diminishes the proportion of air; it is advantageous, however, in causing a more steady flame. But the draught of air is greatly increased if the chimney

Fig. 20. Fig. 21. Fig. 22.

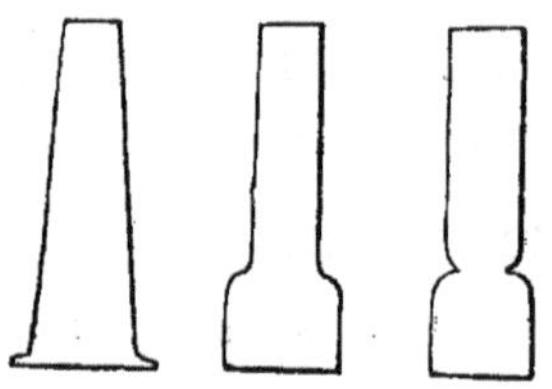

is tall and narrow, or contracted towards the top, as figs. 20, 21; or if it have a sudden contraction near the bottom, as in fig. 22. A short chimney with a slight contraction gives the flame a much better light than a wide chimney does, but the gas (or oil) is consumed faster in just the same proportion. Such a chimney is advantageous where a very strong light is required with a profusion of gas, especially if the gas is unusually rich in carbon, as recently prepared oil-gas. A tangential current of air is produced by the contraction, which sweeps the outer surface of the flame. The same effect is produced at the bottom of the flame by what is called the "double cone gas-burner," which is an Argand burner rising through a double brass cone terminating close to the top of the burner. The air rises between the two cones and impinges on the flame in a tangential current.

Such are the means of increasing the proportion of air brought into contact with the surface of a flame, without diminishing the quantity of gas. But the inconveniences which arise from too much gas, or too little air, are not greater than those which proceed, on the other hand, from too great a supply of air. With a very great draught two evils are likely to result: 1°, the intermixture of air with the interior of the flame, which always happens in such a case to a greater or less extent, produces a more complete combustion of the carbon than is proper; and 2°, a rapid current of air in larger quantity than is necessary for the combustion of the gas, carries away the heat, and thus prevents the flame from attaining its maximum temperature. If a gas-flame from a large Argand burner of several holes be depressed (by turning the stop-cock of the jet) until the yellow flame disappears, the flame which remains is blue, and so feebly luminous as to be hardly visible in broad day-light. But the same or a smaller quantity of gas when made to pass through three or four holes, as by shutting off the remainder of the holes with a flat glass plate, will give a far more luminous flame. In the first case, the proportion of air is so large as to cause complete combustion of the carbon and hydrogen at the same time, but not so in the latter case. When a jet of gas issues from a small orifice with great velocity, it becomes mixed by expansion with atmospheric air, and the deposit of carbon is consequently deficient. A great loss of light is also experienced with a very tall chimney, on account of the rapidity and completeness of the combustion through the unnecessary draught.

Of late years gas has come into very general use in the chemical laboratory as a source of heat. The most convenient mode of applying it for this purpose is in the form of the inflammable mixture with air. The mixture affords a very intense heat, because a considerable quantity of gas is consumed within a small compass, and it has the advantage of giving a flame without smoke. The apparatus represented in figure 23 is found advantageous for this purpose. The mixture of gas and air is made in the chimney;

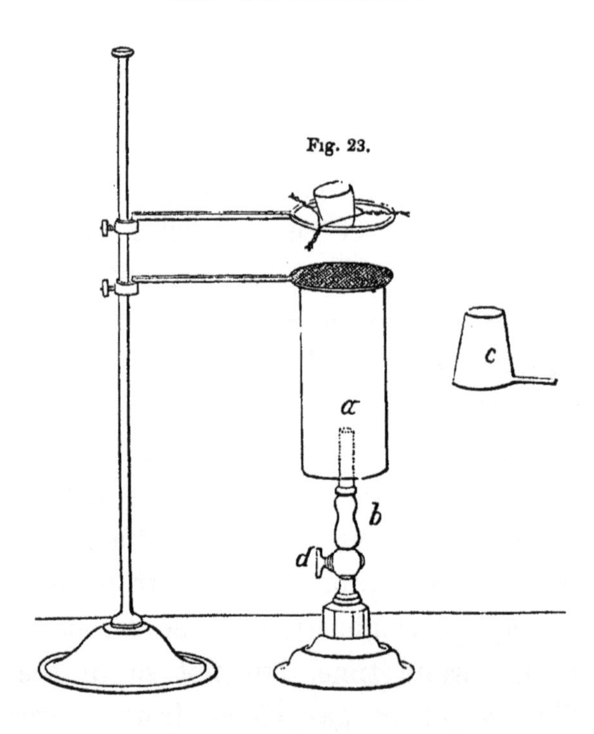

Fig. 23.

a piece of wire-gauze is fastened over the top, above which the inflammable mixture is ignited: *a* is a copper chimney supported by the ring of a retort-stand having a ridge or pegs for that purpose on the outside; the bottom is open to admit air: *b* is a plain gas jet, the extremity of which is near the bottom of the chimney; the crucible or capsule to be ignited is supported over the flame on a triangular piece of iron wire, placed on another ring of the retort-stand: *c* is a second chimney or jacket to be placed over the crucible to create a constant and equal draught. The proportion of gas and air is regulated by the stop-cock *d*, and by raising or depressing the chimney over the extremity of the gas jet. The flame should not burn yellow, but with a clear blue colour, and ought not to produce any deposit of carbon on a body held in it. A little experience will shew what kind of flame produces the most intense heat.

Another excellent mode of obtaining heat through the combustion of gas is by adapting to an ordinary Argand burner a chimney with a contraction, similar to fig. 22, made by simply fastening a circular disc of copper with a round hole in the centre into a common cylindrical copper chimney. It is said that the temperature is further increased by punching out a circle of small holes in the disc, as in the top of an Argand burner.

The attention of the public has been lately drawn, by Mr. Faraday, to the necessity of paying more than ordinary attention to the proper ventilation of gas-lights. The importance of this subject will be readily admitted when it is considered that one part by weight of well-made coal-gas produces in combustion nearly two and a half parts by weight of *carbonic acid gas.* If attempted to be breathed in a state of moderate dilution with air,* that gas soon proves fatal; and causes asphyxia, even if largely diluted. Though the atmosphere of a chamber would hardly acquire from gas-lights the quantity necessary to produce such effects, yet the continued respiration, for several hours, of air containing not more than one or two per cent. of carbonic acid, is known to have produced alarming effects. But carbonic acid gas is not the only deleterious product of the combustion of coal-gas. Unfortunately, none of the ordinary operations

* If pure carbonic acid is attempted to be inspired, a violent spasm of the glottis takes place, which prevents the gas from entering the lungs, and causes suffocation.

of purification will withdraw the whole of the bisulphuret of carbon which is contained in coal-gas, from whatever kind of coal the gas may be made. Bisulphuret of carbon, composed, as its name implies, of carbon and sulphur, is a liquid at common temperatures, but so very volatile that it is retained in the state of vapour by the gas, though exposed to a very low temperature. In the combustion of gas containing this vapour, the sulphur becomes sulphurous acid, which, by the action of moisture and the oxygen in the air, passes into the state of sulphuric acid, or oil of vitriol. This corrosive liquid attaches itself to the walls and furniture of the apartment, and, being very fixed, does not dissipate by evaporation and ventilation. The destruction of the bindings of the books in the library of the Athenæum Club was partly attributed by Mr. Faraday to the action of sulphuric acid thus formed and condensed on the backs of the books. The water collected in a receiver from the combustion of gas at the Athenæum gave abundant evidence of the presence of sulphuric acid when tested by chloride of barium.*

The common mode of carrying away these deleterious vapours, where any means at all are had recourse to, is by suspending at a little height above the chimney of the lamp a bell-shaped vessel, connected at its top with a narrow tube, leading out of the apartment. The diameter of the bottom of the bell-vessel is usually much greater than the diameter of the chimney of the lamp. The ventilation, or the ascent of the hot air from the lamp through the

* Lecture at the Royal Institution, April 7, 1843.

bell-shaped vessel and tube, takes place with a rapidity corresponding to the difference between the density of the hot air within and the external cold air. As the difference in density depends on the difference in temperature, to have a vigorous draught in the tube it is necessary to keep up the temperature of the latter as high as possible. But the extent of surface presented by the bell-vessel allows so much of the heat to escape by radiation, that sometimes the temperature of the tube is scarcely raised above that of the surrounding atmosphere, in which case ventilation hardly takes place at all. This inconvenience may be surmounted (as shewn by Mr. Faraday) by dispensing with the bell-vessel altogether, and conducting a copper tube, of about the same diameter as the flame, from the summit of the flame to the exterior of the apartment. Thus placed, the copper tube necessarily becomes more strongly heated than the bell-vessel in the common mode of ventilation; a rapid current is established; and not only are all the products of the combustion of the gas carried away, but even the external air itself may be sucked in at the top of the chimney. If any objection can be raised to this mode of ventilation, it consists in the rather unsightly appearance of the tube so near the flame.

A very elegant lamp was devised by Mr. Faraday (and patented by his brother), in which the ventilating current is made to descend between two concentric glass chimneys of unequal height, the interior being the lowest, and to pass from the bottom through a tube, which is afterwards bent upwards. The burner is an Argand, supplied with air in the ordinary manner; the tube to conduct away the gases, or ventilating pipe, terminates in a box at the bottom of the lamp, formed of two concentric cylinders, the space between which is closed at bottom, but open at top, in order to communicate with the space between the two chimneys; the Argand burner, with the air necessary to feed it, rises through the interior cylinder of the box at the bottom of the lamp. The exterior chimney, which is the highest, is covered with a plate of mica.

The descending current is determined, in the first place, by applying heat for some time to the bend of the ventilating tube where it begins to ascend; when fully established, the gas is lighted, and the exterior chimney covered with a plate of mica. The gases from the flame then pass from the top of the interior chimney, downwards through the space between that and the exterior chimney, into the box in which terminates the ventilating pipe to convey the gases without the apartment. A globe of ground glass now placed over the lamp gives it a very elegant appearance. It is said that a greater amount of light is obtained with such a lamp from the same quantity of gas, and certainly a larger flame is afforded without opening the stop-cock further; but it seems probable that the force of the draught sucks up a little more gas than would otherwise be emitted. Particular attention must always be paid in this lamp to determine the downward current fairly, by applying heat to the bend of the ventilating tube before the gas is lighted. The neglect of this is sure to be attended with the destruction of some of the glass parts of the apparatus.

§ IX. ECONOMY OF GAS ILLUMINATION.

From the statements contained in the preceding section, it is evident that the amount of light obtained from the combustion of gas, and hence the relative economy of gas-light, depends in no inconsiderable degree on the manner in which the combustion is effected ; and, as the cost of the production of gas entirely depends on the facility of procuring coal, and of disposing of the bye products of the manufacture, which must vary more or less in every locality, it is impossible to make an estimate of the advantage of gas illumination in an economical point of view of general application. The following considerations on this subject have especial reference to the London gas-works.

The average cost of a ton of Newcastle coal, delivered at the works in London, may be taken at seventeen shillings, and the value of the products from the same quantity of coals, as follows :

	£.	s.	d.
8,500 cubic feet of gas at 9*s.* per 1000 cubic feet	3	16	6
36 bushels of coke	0	12	0
12 imperial gallons of ammoniacal liquor	0	0	3¼
190 pounds of tar	0	4	0
	4	12	9¼

According to this estimate there is yielded for every ton of coals about 3*l.* 15*s.* 9*d.*, to be placed against the current expenses of the manufacture and the capital. The fuel employed to heat the retorts in London is coke, of which there is required a little more than one-third of the amount produced ;* therefore

* A ton of coals requires about thirteen bushels of coke (heap measure).

four shillings and threepence may be deducted from the above estimate for fuel. About two-thirds of a bushel of lime, which costs, say ninepence, is required to purify the gas from a ton of coals. Where alum, green vitriol, sulphate of manganese, or sulphuric acid is employed to separate ammonia, the value of the ammoniacal salt formed is a little more than the original cost of the material. After making the above deductions, there remains 3*l.* 10*s.* 9*d.* per ton to be placed against the wages of workmen, salaries of managers, clerks, and collectors, repairs, and interest of capital.

Such is an outline of the advantages of this manufacture to the producers, where the annual consumption of gas amounts to three thousand millions of cubic feet, and where more than thirty thousand street lamps are supplied at an average price of four pounds per lamp per annum. In small establishments the profits are considerably less. It appears that, in most localities in the South of England, where about one hundred lights are required, a coal-gas apparatus may be found profitable; but the recent improvements in the construction of oil-lamps render it doubtful whether a gas apparatus would be advantageous for less than a hundred lights.

With reference to the economy of gas illumination to the consumer, Mr. Peckston offers the following considerations (Practical Treatise on Gas-lighting, 3rd edit. p. 27). An imperial gallon of sperm-oil burned in an Argand lamp, which yields light equal to five candles, will burn about 100 hours, or give an amount of light equal to 500 candles burning one hour each; the expense per hour, when sperm-oil is nine shillings per gallon, will be a little more than a penny for such a lamp; but when whale or seed oil, at three shillings, or three shillings and sixpence per gallon, is substituted for sperm-oil, and the solar lamp used, which furnishes light equal to 4¾ mould candles of six to the pound, an imperial gallon will burn about 90 hours, and give a total amount of light equal to 427½ candles, each burning one hour. The solar lamp with whale or seed oil costs something less than a halfpenny per hour. It is generally admitted that an Argand burner of fifteen holes consumes about five cubic feet of gas per hour, and yields light equal to twelve mould candles of six to the pound.

Taking as a standard the amount of light which would be produced by a certain number of wax candles of six to the pound, the relative cost of that amount of light from other sources is the following:

Wax candles	100·0
Sperm-oil burned in Argand lamps	49·6
Ill-snuffed tallow candles	49·6
Tallow candles (mould) six to the pound	36·8
Whale or seed oil burned in the "overflowing shadowless lamp," or the "solar lamp"	23·4
Coal-gas	10·0

Or, if the cost of wax candles is estimated at one shilling, the cost of the other sources of equal light will be as follows:

		£.	s.	d.
Wax		0	1	0
Sperm-oil		0	0	6
Ill-snuffed tallow candles		0	0	6
Tallow candles		0	0	4½
Whale or seed oil		0	0	2¾
Coal-gas	nearly	0	0	1¼

The following table, shewing the relative intensity and cost of light obtained from different sources, was constructed by Mr. Rutter, engineer to the Old Brighton Gas Company. The cost of gas is estimated at ten shillings per 1000 cubic feet, and the numbers in the third column represent the number of cubic feet of gas which afford as much light as the quantities of the materials mentioned in the first column.

1. Source of light.	2. Cost.		3. Cubic feet of gas.	4. Cost of gas.	
	s.	*d.*		*s.*	*d.*
1 ℔ tallow candles (moulds)	0	8	21	0	2½
1 ℔ composition candles	1	0	25	0	3
1 ℔ wax candles	2	6	25	0	3
1 gal. whale (solar) oil	3	6	175	1	9
1 gal. sperm-oil	9	0	217	2	2

The table in the following page is a copy of a paper laid before a Committee of the House of Commons by Mr. Hedley of the Alliance Gas Works, Dublin, shewing the relative economy of the gas manufactured at different places in this kingdom, compared with candles. The price of one hundred pounds of the candles referred to is 3*l.* 2*s.* 6*d.*, and that quantity is estimated to burn, one at a time, for five thousand seven hundred hours.

TABLE OF THE RELATIVE VALUE OF COAL-GAS MADE AT VARIOUS WORKS, COMPARED WITH CANDLES.

(BY MR. HEDLEY.)

Names of the places.	Illuminating power of a single jet of gas-flame, four inches high, taken by a comparison of shadows.	The jet of gas burnt, four inches high, consumed per hour, and was equal to the candles in the preceding column.	Gas required to be equal to 100 pounds of mould candles, 6 to the pound, each 9 inches long.	Selling price of gas per 1000 cubic feet according to metre.	Cost of gas equal in illuminating power to 100 pounds of candles.	Average discount allowed off the charge for gas.	Net cost of gas equal to 100 pounds of candles.	Specific gravity of the gas.
	Equal to candles.	*Cubic feet.*	*Cubic feet.*	*s. d.*	*£. s. d.*	*Per cent.*	*£. s. d.*	
Birmingham; Birmingham and Staffordshire (two companies).	2·572	1·22	2704	10 0	1 7 0	9	1 4 7	·541
Stockport	3·254	·85	1489	10 0	0 14 11	12½	0 13 0	·539
Manchester	3·060	·825	1536	8 0	0 12 3	11¼	0 10 10	·534
Liverpool New Gas Company	4·408	·9	1164	10 0*	0 11 8	6¼	0 9 10	·580
Bradford	2·190	1·2	3123	9 0	1 8 1	12½	1 4 6	·420
Leeds	2·970	·855	1644	8 0	0 13 2	6¼	0 12 4	·530
Sheffield	2·434	1·04	2440	8 0	0 19 6	6¼	0 18 3	·466
Leicester	2·435	1·1	2575	7 6	0 19 3	15	0 16 5	·528
Nottingham	1·645	1·3	4200	9 0	1 17 9	15	1 11 3	·424
Derby	1·937	1·2	3521	10 0	1 15 4	15	1 10 0	·448
Preston	2·136	1·15	3069	10 0	1 10 8	15	1 6 2	·419
London	2·083	1·13	3092	10 0	1 10 11	none allowed.	1 10 11	·412

* Since this table was compiled the price of gas from the Liverpool New Company has been reduced to seven shillings per 1000 feet.

Concerning the relative economy of coal-gas and oil-gas as sources of light, no estimate can be offered of constant application, as the price of oils, from the precarious nature of the sources from which they are obtained, is subject to considerable fluctuation. If the relative cost of light from coal-gas and oil-gas is estimated as one for the former, to five or six for the latter, the cost for oil-gas is probably under the truth. But the illuminating power of oil-gas is far greater than that of coal-gas, taking bulk for bulk. Recently prepared oil-gas has occasionally more than three times as much illuminating power as coal-gas; the average is about the proportion of 1 of coal-gas to 2·6 or 2·7 of oil-gas. The following table by Drs. Christison and Turner shews the relative illuminating powers of these gases at different densities:

	Density.		*Proportion of light.*	
	Coal-gas.	*Oil-gas.*	*Coal-gas.*	*Oil-gas.*
	659	818	100	140
	578	910	100	225
	605	1110	100	250
	407	940	100	354
	429	965	100	356
Mean	535	948	100	265

Of oil-gas of the specific gravity of 900, an Argand burner giving a light equal to seven mould candles consumes a cubic foot and a half per hour.

§ X. MODES OF ESTIMATING THE ILLUMINATING POWER AND PURITY OF LIGHT-GAS.

The only method of strict accuracy by which the value of an illuminating gas may be determined is by a complete chemical analysis; but this is an operation of extreme delicacy, requiring far more adroitness in minute manipulation than is possessed by the generality of gas engineers. Results, however, of sufficient accuracy for all ordinary purposes may be readily obtained by methods much more easily executed, namely, 1°, by a photometrical experiment; 2°, by determining the specific gravity of the gas; and, 3°, by determining the quantity of oxygen required for the complete combustion of the gas, with the amount of carbonic acid produced. The first of these methods is the simplest; and the results it affords, when performed with care, are equal in value to those got by the two other methods.

The photometrical processes which I shall briefly describe are founded upon principles of extreme simplicity. Though the eye is unable to judge with precision of the relative intensity of two lights, yet it can determine with considerable accuracy when contiguous shadows of an opaque object thrown upon a screen by different lights are equally dark, or when two similar adjoining surfaces are equally illumined, provided the lights are of the same tint. If the two lights which produce these effects are equal in intensity, obviously their distance from the screen must also be equal; but if unequal, the most intense light is placed farthest from the screen. Now, as rays of light are propagated continually in straight divergent lines, their intensity diminishes in the direct proportion of the square of their distance from their source. Taking as a standard, the amount of light on a screen derived from a flame at the distance of one foot, then at two feet the light on the screen from the same source would be one-fourth, at three feet one-ninth, and at four feet one-sixteenth of the standard. Therefore, if two or more sources of light are so placed as to cast an equal light on the screen, their relative intensities are directly as the square of their distances from the screen. The objection to this mode is, that it does not readily admit of a fixed standard of comparison.

The method of contrasting the shadow of an opaque object formed by different lights was first employed by Lambert (*Photometria*, 1760), but is commonly attributed to Count Rumford, by whom it was proposed in the Phil. Trans., vol. lxxxiv. The apparatus required is extremely simple, consisting merely of a smooth perpendicular surface of uniform colour, and a rod for throwing the shadow. The two lights which are to be compared are so placed, that, when the rod is interposed between them and the screen, the two shadows may be contiguous; and, so long as the shadows are of unequal depth, one of the lights must be advanced towards, or retired from, the screen, until an equality in depth is procured. Suppose a wax candle at the distance of two feet, and a gas jet at the distance of two feet six inches, to produce equal shadows, then, according to the above rule, the relative intensity of the lights is as 4 to 6·25, or as 1 to 1·5625.

Similar in principle is the elegant little instrument constructed for the same purpose by the late Dr. Ritchie, by which a very correct estimate may be made of the relative brightness of two lights, provided they are of the same tint. The apparatus, which is shewn in section lengthways in figure 24, consists of a rectangular box about two inches square, open at both ends, and blackened upon its inner surface to absorb extraneous light. On the top of the box is a narrow slit about one inch long, and one-eighth of an inch broad, covered with tissue or oiled paper. Within are placed two rectangular plates, *a*, *b*, of plane looking-glass, cut from the same piece to ensure uniformity in reflecting power. Their width is the same as that of the box, and their length may be equal to the hypothenuse of a right-angled isosceles triangle, whose side is the height of the box, or a little less, as in the above figure. The plates are fastened together so as to meet at the top in the middle of the slit (or in the line of that perpendicular), their reflecting surfaces being towards the open ends of the box. In using this instrument, it is placed in a straight line between the two lights whose intensities are to be compared, so that the light from each source is reflected from the respective mirrors to the tissue. The instrument is then moved nearer one or the other light, until, to an eye situated above the box, the two portions of the slit which correspond to the respective mirrors appear equally illuminated. The squares of the distances of the lights from the vertical, *c*, give the proportion of the intensities required.

Fig. 24.

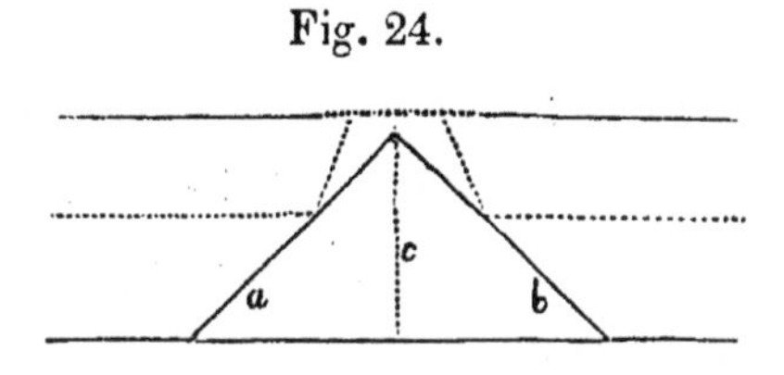

A very ingenious and accurate mode of determining when two rays of light from different sources have the same intensity, has been proposed by M. Arago; but this method will hardly supersede those just described, at least for the use of gas-engineers, until a more simple apparatus is devised for its application. The method of M. Arago is founded on the property possessed by a ray of light of dividing itself, when polarized and passed through a doubly refracting prism, into two rays of the complementary colours, red and green. Rays from the two lights which are to be compared are polarized by the ordinary means, as transmission through a plate of tourmaline, or reflection from glass at the polarizing angle, then received on a plate of rock crystal, and observed through a doubly refracting prism. Each light will then give two images tinged with red and green. On bringing the images into such a position that the red of one light falls over the green of the other, if the intensities of the two lights are equal, the superposition completely neutralizes the colour, and a white or colourless image is the result; but, if the intensities are unequal, the image is slightly coloured with red or green, according as the one or the other predominates.

Another mode of estimating the illuminating power of coal-gas is by determining its density, the light afforded by illuminating gases being in general proportioned to their density. The most valuable constituent of coal-gas, namely, olefiant gas, has the density 997 (air as 1000); the next in value, light carburetted hydrogen, has the density 560; and the density of hydrogen, which is the principal deteriorating constituent, is 69. Thus far it would seem that the illuminating power of coal-gas might be estimated with accuracy by taking its density; but, unfortunately, carbonic oxide and nitrogen, which are of no value as sources of light, are heavy gases compared with light carburetted hydrogen and hydrogen: the density of carbonic oxide is 967, and that of nitrogen 970. Hence the estimate thus obtained cannot be relied on in all cases. The density of the gas may form, nevertheless, a valuable datum in the estimate, as appears from the following results of some experiments by Mr. Hedley:*

Density of gas.		*Number of candles equal to gas when burned in a single jet four inches high.*
412	two specimens	1·562
412		1·562
420		1·645
424		1·234
448		1·453
453		1·929
455		1·929
462		1·777
466		1·826
528		1·826
530		2·228
534		2·295
539		2·441
580		3·306

The late Dr. Henry, whose researches on coal-gas form the subject of memoirs of great interest, considered that the comparative value of the different light-gases may be accurately determined by the quantity of oxygen required for the perfect combus-

* For a short account of the method of obtaining the density of gases, see "Elements of Chemical Analysis," p. 274; and, for more minute details, consult Faraday's "Chemical Manipulation."

tion of equal volumes, and the quantity of carbonic acid produced thereby. The larger the proportion of hydrocarbons present, the greater amount of oxygen will be required ; light carburetted hydrogen requires more oxygen for combustion than carbonic oxide or hydrogen, and olefiant gas more than light carburetted hydrogen. In fact, both the illuminating powers of the different gases, and the amount of oxygen required for their combustion, depend in a great measure on the proportion of gaseous carbon contained in one volume of the gas. If one hundred volumes of one gas require for perfect combustion one hundred volumes of oxygen, and one hundred volumes of another gas require two hundred volumes of oxygen, the value of the second will be double, or a little more than double that of the first.

A convenient instrument in which to cause the combination of the gas with oxygen is Dr. Ure's Eudiometer, represented in fig. 25. It is formed of a stout glass tube about twenty inches in length, and a quarter of an inch internal diameter, sealed at one end and doubly bent at the middle. The limb with the sealed end is graduated into equal parts according to an arbitrary scale. Two platinum wires are sealed into the glass near the closed end, with their extremities within the tube about one-tenth of an inch apart. On placing one of these wires in communi-

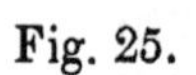
Fig. 25.

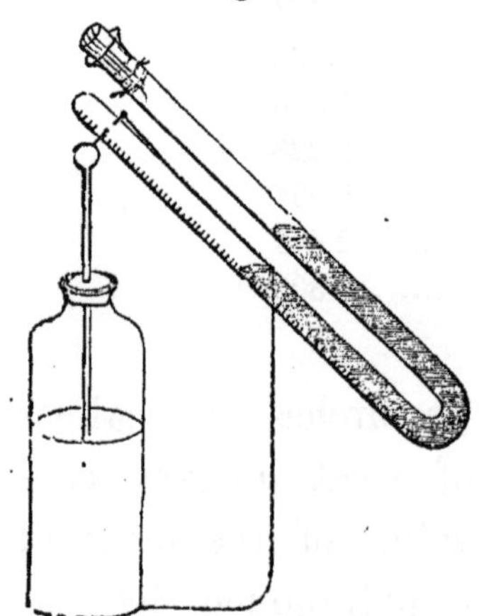

cation with the knob of a charged Leyden jar, and the other with the outside of the jar or the ground by a metallic chain, the electric spark passes within the tube. In the first place, a mixture of the gas to be tested, and pure oxygen, in known proportions, the oxygen being in excess (say, three volumes of oxygen to one volume of the gas), is made in a graduated jar at the water-trough. Enough of this mixture is transferred to the eudiometer, previously filled with water or mercury, to occupy about two inches of the sealed end; and the amount of the mixture introduced is carefully observed, the liquid standing at the same level in both limbs. The mouth of the open limb is now closed with a cork, which is secured by a wire, and the electric spark taken through the wires. The gaseous mixture explodes, and immediately afterwards a contraction is perceived from the condensation of the steam formed by the combustion. The gas which remains, supposing the combustion to be complete, consists of the excess of oxygen, carbonic acid formed by the combustion, and the nitrogen in the coal-gas. The liquid is again brought to the same level in both limbs, and the measure of the gas is accurately noted: the carbonic acid gas is next absorbed by agitating the gas with a little solution of caustic potash; for which purpose it is convenient to introduce a few fragments of fused potash through the open end of the tube, and shake them into the sealed limb. With a slight agitation the whole of the carbonic acid is immediately absorbed. The measure of the remaining gas being noted, and the closed end of the instrument completely filled with water, by a little dexterity the gas is brought into the limb with the open end. The instrument being inverted (the bend upwards), and the open end under water at the pneumatic trough, a stick of phosphorus attached to a piece of copper wire is introduced into the open limb and brought into contact with the gas. The phosphorus soon absorbs the excess of oxygen in the mixture, and what remains is the nitrogen of the coal-gas, to determine the amount of which it must be returned to the sealed limb. The excess of oxygen, and therefore the quantity required for the combustion, is also then ascertained. The results obtained should always be tested by a second similar experiment; and by a third, if any discrepancy is observed in the first two results.

From the data obtained by such an experiment, that is, the volume of oxygen employed in the combustion, and the volume of carbonic acid produced, a tolerably accurate estimate of the value of illuminating gases may be formed. The reason of this will appear on examining the result of the combustion with pure oxygen of each of the combustible gases contained in coal-gas :

Name of gas.	*100 vols. of gas require of oxygen*	*Contraction of the mixture after explosion.*	*Vols. of carbonic acid produced.*
Olefiant gas . .	300 vols.	one-half	200
Light carburetted hydrogen . .	200 „	two-thirds	100
Hydrogen . .	50 „	complete	——
Carbonic oxide .	50 „	one-third	100

That a relation generally subsists between the amount of oxygen required, and the value of the coal-gas, may be perceived by the table at page 81, shewing the composition of coal-gas evolved at different periods of the distillation. See also the table of the composition of oil-gas, page 99.

For the complete analysis of coal-gas, the process given in the following table seems to be the best we are possessed of, but it is not altogether free from objections. The chief source of error lies in the determination of the carbonic oxide and hydrogen, for potassium, at a particular temperature, absorbs some hydrogen as well as carbonic oxide; but this circumstance does not interfere with the estimation of the important constituents of coal-gas.

ANALYSIS OF COAL-GAS.

The constituents to be estimated are carbonic acid, olefiant gas and vapours of hydrocarbons, carbonic oxide, light carburetted hydrogen, hydrogen, and nitrogen.

Agitate the gas in a graduated jar with a solution of caustic potash.

Carbonic acid is absorbed.

Transfer the remaining gas to another graduated jar, of about half an inch in diameter, standing over water; mix it with half its bulk of pure chlorine, and keep the mixture carefully protected from the solar rays.

Olefiant gas and *vapours of hydrocarbons* are absorbed. The absorption is complete in the course of 24 hours.
Remove the excess of chlorine from the remaining gas by agitation with a solution of caustic potash, and note the bulk of the residue.

Heat potassium gently in some of the remaining gas contained in a recurved tube over mercury.*

Carbonic oxide is absorbed.

Transfer a portion of the remainder into the eudiometer, and introduce a known measure of pure oxygen (not less than twice the volume of the gas), and detonate by the electric spark. Note the quantity of gas, and introduce a solution of caustic potash.

Carbonic acid equal in bulk to the quantity of *light carburetted hydrogen* is absorbed.

The residue consists of nitrogen and the excess of oxygen. Introduce phosphorus to absorb oxygen, that the *amount of oxygen employed in the combustion* may be found. From this deduct twice the volume obtained of light carburetted hydrogen; twice the bulk of the remaining oxygen represents that of the *hydrogen*. The residuary gas is *nitrogen*.

* This operation is best performed in a glass tube, closed at one end, about eight inches in length, and three-quarters of an inch in diameter, with one inch and a half from its top recurved to hold the potassium.

In the preceding method of analyzing coal-gas, the olefiant gas and vapours of hydrocarbon are estimated together as the contraction by chlorine. The determination of the hydro-carburetted vapours may be accurately effected, as shewn by Mr. Faraday, by means of oil of vitriol. That liquid absorbs both the vapours and olefiant gas, but the latter not so rapidly as the former; and if the coal-gas is diluted with three or four times its volume of atmospheric air or hydrogen, and the mixture kept in the shade, the absorption of olefiant gas is prevented. Another portion of the gas under examination may be taken for the remaining operations.

As the principal luminiferous constituents of light-gas are olefiant gas and the vapours of hydrocarbons, a good approximation to its true value may be obtained by determining merely the contraction which the gas experiences when mixed with chlorine, neglecting the other operations in the preceding method of analysis; and instead of removing the excess of chlorine by caustic potash at the close of the experiment, the amount of chlorine absorbed by water in similar circumstances may be noted and deducted from the entire condensation. The necessity of carefully excluding the mixture from solar light, arises from the property which chlorine possesses of condensing carbonic oxide under the influence of solar light and moisture; but no condensation from the presence of carbonic oxide takes place in the dark, or in candle-light. The purity of the chlorine employed in this process should be previously ensured by ascertaining whether it is absorbed by water, or by a dilute alkaline solution without leaving a residue. The chlorine should be added to the gas so long as the volume of the latter diminishes.

§ XI. REGULATORS AND METERS.

The flow of gas from the gasometer for distribution along the pipes is not commonly regulated by a stop-cock, but by a water or mercurial valve, which is, in fact, merely a small and very delicate gasometer. The water-valve has precisely the construction of the ordinary gasometer, excepting being square, and the upper part having a vertical partition descending about half-way from the covered top to the bottom; not so far as the surface of water in the cistern, except when it is required to cut off the supply of gas entirely. This partition is placed between the entrance and exit pipes, so that, in proportion as the weight of the counterpoise is diminished or increased, the partition approaches or recedes from the surface of water in the cistern, and of course retards or permits the flow of gas from the entrance to the exit pipe. The flow of gas is, therefore, regulated by the weight of the counterpoise. With this small gasometer is sometimes connected a vertical rod, which carries a black-lead pencil made to bear upon a paper cylinder; the latter rotates on its axis by communication with a time-piece. By this contrivance every change of pressure which may take place during the absence of the observer is shewn by the aberration of the line; it is, therefore, a pressure register.

A sufficiently precise idea of the construction of the mercurial valve may be formed by the consideration of fig. 26. In this apparatus the cistern is the moveable part to regulate the flow, the upper part being fixed; *a* and *b* represent the two gas-pipes terminating in a rectangular iron vessel *c*, open at bottom, which dips into the cistern *d*, containing mercury up to the level *e*. By means of a screw working against the bottom of the cistern, the latter is raised or lowered so as to bring the surface of mercury nearer to, or to draw it farther from, the vertical partition *f*, by which the flow of gas is impeded or retarded at pleasure. The screw may be turned by an index working over a graduated circle, by which the exact size of the opening between the bottom of the partition and the surface of mercury may be indicated.

Fig. 26.

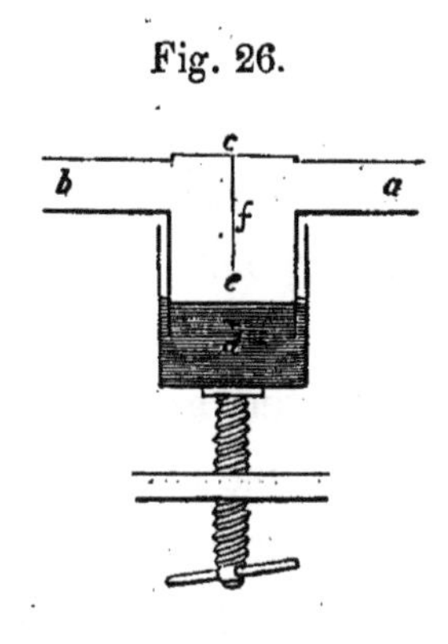

In large gas-works a regulator of a more complicated construction is made use of, so contrived as to preserve at all times a given rapidity in the passage of gas, however fluctuating the pressure of the gasometer. The principle of the action of this instrument, different modifications of which are made use of, may be understood by reference to fig. 27. It consists of a rectangular cistern containing water up to the level *a a*; *b*

Fig. 27.

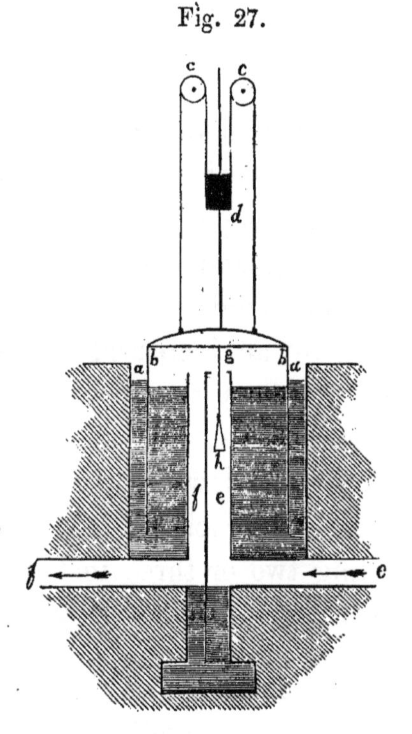

is the part which answers to the gas-holder of the ordinary gasometer, sustained at any required height by chains working over the pulleys *c c*, and bearing the counterpoise *d*. The gas enters by the pipe *e*, and passes out by *f*. Within the entrance pipe, which is contracted at top, is suspended by the chain *g*, a cone *h*, the base of which is a little wider than the aperture at the top of the pipe. Whenever the pressure on the gasometer is increased, the weight of the counterpoise *d* remaining the same, the vessel *b* rises, lifts the cone, and contracts the aperture through which the gas enters. By properly adjusting the length of the chain *g*, the flow may be at all times equally maintained, having been first adjusted by the counterpoises. The pressure in the gasometer may be ascertained and registered as in the common water-valves.

Platow's "gas moderator" is an ingenious instrument for preserving an uniform flow of gas on the small scale. The extremity of the tube which supplies the gas is immediately opposed by a disc, which is connected with a spring and crank in such a manner, that when the former would be driven from the extremity of the gas-pipe by an increased flow of gas, if not subject to the action of the spring, the latter causes the disc to approach the tube, and therefore offer an impediment to the flow of the gas. These parts of the moderator are enclosed in an air-tight cylinder hardly exceeding two or three inches in length, from the top of which issues the gas delivery tube.

The consumers of light-gas in London and many other places are now generally required by the respective gas companies to burn the gas by measure. The principle of the action of the ordinary *gas meter*, or apparatus for measuring the quantity of gas which passes through a pipe, may be understood with the assistance of fig. 28, which represents a section of the working part of the apparatus perpendicular to its axis. This part consists of a cylindrical case *a*, within which revolves a shorter cylinder *b*, shut at both ends, and divided into four compartments by the partitions *c*, *d*, *e*, *f*; each compartment communicates with the exterior vessel by slits on the periphery of the cylinder. The size of the diameter of the cylinder of a meter to supply five lights is about thirteen inches. About two-thirds of the cylinder revolve in water, and the gas enters the space above by the elbow tube *g*, which enters at the axis and terminates immediately above the surface of the water. As the gas enters, and exerts a pressure against the partition *c*, it causes the cylinder to revolve on its axis from left to right; and, when the slit *h* gets above the surface of the water, the gas in that compartment escapes into the exterior cylinder, from whence it is conducted away by a tube not shewn in the figure. Connected with the opposite end of the axis of the cylinder is an endless screw, which moves a toothed wheel attached to an upright shaft, and the latter communicates with the dial-work of the registers, by which the number of cubic feet of gas which pass through the apparatus in a given time can be read off.*

Fig. 28.

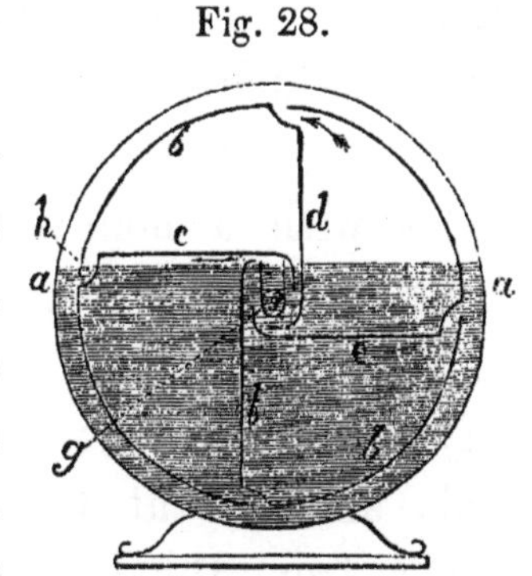

* Considerable sensation has been lately excited among the gas consumers of the metropolis in consequence of the allegations contained in a

Clegg's "Patent Dry Gas Meter" is an ingenious apparatus, intended to shew the quantity of gas which passes through it, by measuring the temperature of a little brass cylinder heated by an inflamed jet of the gas. The jet proceeds from the side of a small brass knob about an inch and a half long, the upper part of which is without the meter. The bottom of this knob, within the meter, is connected at right angles with a hollow brass case, called the *heater*, about two inches long, and half an inch broad. The gas enters the meter at a prolongation of one extremity of the heater; a minute quantity passes up the knob, and out at the jet, but the bulk of the gas escapes from the heater downwards into the body of the meter by three descending tubes, each about three-quarters of an inch in length. By its passage through the heater and the three descending tubes, the gas becomes heated; a measure of its temperature would, therefore, other circumstances being equal, be a measure of the intensity of the combustion in the small flame issuing from the knob at top.

The manner in which the temperature is measured and the registration effected, may be understood by reference to fig. 29 : *a* is the cone, the dot in which

pamphlet on gas meters by Mr. H. Flower, in which the author proves that the registry of the ordinary gas meter is subject to an excess over the actual amount of gas transmitted amounting to from eight to thirty per cent. The chief source of error seems to consist in the want of a proper adjustment of the surface of the water; an undue quantity of water in a meter, according to Mr. Flower, will make an excess of about twenty per cent. in the amount registered over the quantity actually consumed. In Edge's "Improved gas meter" an excess of water is prevented by having an open tube or waste-pipe leading from the proper surface of the water downwards into a close cistern, or waste-water box, from which it is withdrawn when requisite by an opening at bottom.

represents the aperture through which the jet of gas issues; *b* is the heater in cross-section; *c* one of the tubes descending from the heater, by which the gas enters the body of the meter; *d* is a glass instrument similarly constructed to the differential thermometer of Leslie; but *e* and *f*, instead of being two round bulbs, are two cylinders of three inches in length. This thermometer, the cylinders of which are half filled with alcohol and exhausted of air, is made to vibrate by the rod *g* on the centre *h*, and is balanced by the weight *i*. Supposing the apparatus in action, and the thermometer placed as in the figure, the heated gas descends through the three tubes, of which *c* is one, and impinges on the top surface of the cylinder *f*. The vapour of alcohol in this cylinder is immediately expanded, and drives the liquid into the cylinder *e*; the latter soon becomes the heavier, and descends until it lies immediately under the tube *c*, when it becomes heated in its turn and again rises. A pendulous motion is thus kept up so long as the gas flows, and the jet in the knob continues burning; and each vibration, being communicated to a train of wheel-work, is registered in the ordinary manner. The number of vibrations in a given time corresponds with the intensity of combustion in the small jet of flame from the knob, which is nearly, though not exactly, proportional to the quantity of gas which passes through the meter.

Fig. 29.

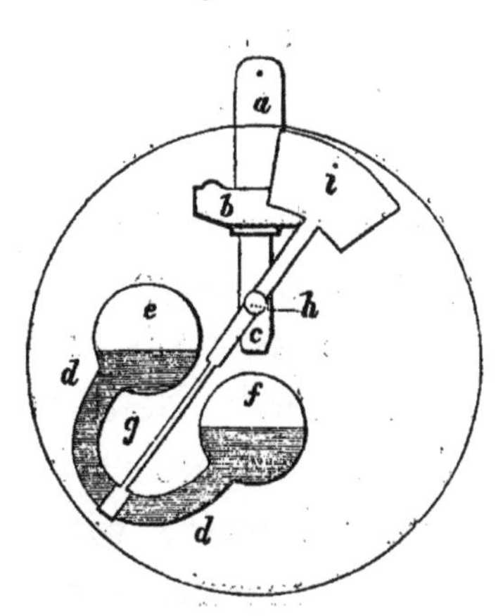

Cast-iron hoods, not represented in the figure, project over the upper glass cylinder on each side, which serve to carry off any superfluous caloric, so that the temperature of the heater and that of the hood which surmounts the top glass cylinder may bear the same relative proportion to each other, whatever the temperature of the external atmosphere. A meter of this construction, of about five inches in diameter and four inches deep, is said to be sufficiently large to measure gas for six burners, and is called a "six-light meter." Neither the temperature nor pressure of gas from the main affects the indications of this meter; it works without membranes or valves, and without interfering with the steadiness of the light: it is cheaper than the common gas-meter, and is not subject to the great irregularities of the latter from variations in the water-line. But, unfortunately for the accuracy of the registrations of this instrument, the temperature of a gas-flame is not proportional to its light: as a measurer of the temperature of the flame, and also of the light of a gas-flame if gas of the same quality is always used, it might probably be made to afford very accurate indications, but there is reason to doubt its accuracy as a measurer of the light of gas of varying quality. Taking the average, however, of a periodical consumption, it may probably be made as accurate as most other meters.

A modification of the ordinary meter has been contrived by Mr. Lowe, by which all pressure of the gasometer may be dispensed with. The chief peculiarity in its construction is, that it possesses the action both of a suction-pump and a forcing-pump; by the former it draws the gas from the service-main, without assistance from the gasometer, and by the latter it emits the gas for combustion. In the place of the water in which the drum of the common meter revolves, a solution of caustic alkali is substituted, which does not freeze as readily as water, and, moreover, serves to separate from the gas any sulphuretted hydrogen, carbonic acid, and sulphurous acid yet remaining.

A large meter, called the *station meter*, is placed at the gas-works between the purifier and the gasometers, to ascertain at pleasure the quantity of gas made during any given period, so that the weekly, monthly, and annual production can be compared with the quantities registered by the meters of the consumers.

Naphthalized gas.—One of the most important of the late improvements in gas illumination is Mr. Lowe's process for naphthalizing gas, or for impregnating ordinary coal-gas with the vapour of naphtha (pages 60 and 94) almost immediately before it issues from the burner. To effect this, the coal-gas is passed either through some porous substance, as sponge or pumice-stone saturated with naphtha, or else over surfaces of that liquid contained in shallow trays. For the first method a rectangular box is employed, called the *naphtha-box*, made of tin-plate or other suitable material, divided into two compartments by a vertical partition in the middle, but which does not quite reach to the bottom of the box. Each compartment contains a series of shelves formed of wire gratings, on which are placed strata of sponge or pumice-stone saturated with naphtha. The gas enters the box by an opening in the end at top, passes down through the several shelves of one compartment, then under the partition and up through the shelves of the other compartment, escaping from the box through an opening corresponding to that at which it entered.

The apparatus employed to saturate the gas with vapour, by passing it over surfaces of the liquid, is equally simple. A rectangular box is fitted up with five or six shallow trays to contain the naphtha, placed one over the other. Near the end of each tray is a slit or aperture extending across the whole breadth of the tray, and furnished with sides rising to about half the depth of the tray. These apertures, which are placed alternately at opposite ends of contiguous trays, answer the purpose of waste-pipes to conduct the naphtha from one tray to that next below it, when the liquid in the former rises higher than the side of the opening. The naphtha is poured by a funnel into the uppermost tray, which it fills to the top of the waste aperture, and then, falling into the trays below, fills them also in succession. When the naphtha issues from a small stop-cock at the bottom of the box, a sufficient quantity has been introduced. The gas enters the box at bottom from the meter, and, passing upwards through the several apertures, is exposed to the surface of the naphtha in each tray successively. By the time it reaches the top, it is thoroughly saturated with the vapour, and is then emitted for combustion.

The latter plan is more convenient than the former

in one respect, namely, that it does not require so frequent replenishing with naphtha. A thousand cubic feet of gas absorbs about a gallon of naphtha.

The increase in the illuminating power of coal-gas by being impregnated with naphtha is very considerable. A jet of hydrogen gas, which burns, when pure, with a pale blue flame hardly visible in the direct solar rays, gives a brilliant flame if the gas is passed through naphtha previous to being ignited. A good flame may be obtained even by impregnating common air, moderately warm (as the breath), with naphtha vapour. It has been estimated that one thousand cubic feet of naphthalized gas is equal in illuminating power to two thousand feet of common gas. If such is the case, the saving in this process amounts to twenty-five per cent., taking the cost of a thousand feet of gas at nine shillings. The expense of the naphthalized gas is

	s.	*d.*
1000 cubic feet of gas	9	0
1 gal. of naphtha	3	6
License for using the process per 1000 feet . . .	1	0
Cost of naphthalized gas equal to 2000 feet of common gas	13	6

Saving 4*s.* 6*d.*, or 25 per cent.

But it is doubtful whether the illuminating power of the gas is so much as doubled; according to another estimate, which seems nearer the truth, the saving is about 15 per cent., reckoning the common gas to be burned in the most advantageous manner.

The naphthalizing box, of the construction first described, is made subservient to the separation of any traces of sulphuretted hydrogen and carbonic acid which may have escaped the ordinary purifying processes. To this end the sponge or pumice-stone in the first compartment through which the gas passes is saturated with caustic potash or caustic soda, the shelves of the other compartment containing naphtha as usual. The small quantity of ammoniacal gas, which coal-gas purified in the ordinary manner generally contains, not only deteriorates slightly the illuminating power of the gas, but is injurious to the pipes. To separate this, a third compartment is given to the naphthalizing box, the porous material on the shelves of which contains dilute sulphuric, muriatic, or other suitable acid. The metallic partitions and sides of the box may be protected from the acid by a covering of a mixture of bees-wax and tallow. In a box of three compartments, the first may contain sulphuric acid, the second caustic soda, and the third naphtha.

As the tar produced in the distillation of coal may be regarded as a kind of intermediate product in the conversion of coal into gas, its formation is attended therefore with the loss of a proportion of gas. Many contrivances have been suggested to reduce the amount of the product of tar, or to convert the tar into gas, by exposing the mixture as it issues from the retort to the further action of heat. In 1835 Mr. John Malam, of Hull, obtained a patent for the use of a second retort or "regenerator," as termed by the patentee, into which the gas and tar vapour are passed before being conducted into the hydraulic main. The regenerator is a cylindrical vessel like a retort, but composed of two tubes, one within the other, furnished with mouth-pieces and arranged as a retort. The mixture of gas and vapours passes from the retort by a tube into the interior tube of the regenerator through a prolongation or mouth-piece at one end, passes into the other end of the tube, which is open, and then escapes into the outer cylinder. After having traversed the whole length of the latter, heated to redness, it is conducted to the hydraulic main.

The quantity of tar obtained when the process is thus conducted is said to be only one-third of the ordinary product, the remaining two-thirds being converted into gas. According to some experiments by Mr. Peckston, the average production of gas when Mr. Malam's regenerator is made use of is 12,500 cubic feet per ton of coals. The quality of the gas is not deteriorated, as might at first be supposed, but is improved; the average specific gravity of the gas obtained by Mr. Peckston being 569. It is stated that the quantity of coal required for charging the retorts, and also of fuel for heating them, may be reduced 25 per cent. The Marquis Montaubon and Messrs. Lowe and Kirkham have also obtained patents for apparatus by which the greater portion of the tar is converted into gas, similar in principle to Mr. Malam's regenerator.

(*For some valuable suggestions which the Editor has received in this article, he is greatly indebted to Mr. John Leigh of Manchester, the chemical director of the gas-works of that town.*)

nd of two tubes, one within the other, terminating
h mouth-piece can inserted as a retort. The
ture of gas and vapours passes from the retort
e tube into the interior tube of the regener-
through a prolongation or mouth-piece at one
d, passes into the other end of the tube, which is
n, and then escapes into the outer cylinder. After
ly traversed the whole length of this heated
nduct, it is conducted to the hydraulic main.
The quantity of the oil obtained when this system is
e conducted appears to be only one-third of the
inary product, the remaining two-thirds being con-
ted into gas. According to some experiments
. Peckston, the average production of gas with
. Mallet's regenerator is put at [illegible]
ic feet per ton of coals. The quality of the gas
not determined, as it may be presumed that
improved; the average increase in the gas
tained by the [illegible] per cent.
, the quantity of coal required for distillation
rts, and also of fuel for heating them, may be
duced 25 per cent. The Messrs. Laming and
ssrs. Lowe and Kirkham have likewise had patents
apparatus by which the oily portion of the tar
converted into gas, similar in principle to the
regenerator.

(For some valuable remarks [illegible]
[illegible] in the Athenæum [illegible]
John Leigh, of Manchester, the chemical director
the gas-works of that town.)

PRESERVATION OF WOOD.

§ I. Properties and Composition of Wood.—II. Nature and Causes of Decay of Wood.—III. Preservative Materials.—IV. Modes of Applying Preservative Materials.—V. Other effects of the Impregnation of Foreign Substances.

THE changes to which wood is liable under certain circumstances, known as rot and decay, are strictly chemical changes, and can only be properly understood and guarded against by their chemical investigation.

When wood is exposed to the simultaneous influence of water and air, it often loses all its tenacity, becoming pulverulent and at the same time darker in colour. If a piece of wood is examined which has lain for some time in a damp situation, its surface being partially protected by paint or varnish, it is found that the parts which are unprotected have become corroded, soft to the touch, and probably, in great part, detached from the mass. For some time after the decay is commenced the protected parts remain sound; but, if the decayed portions are not removed, the entire piece becomes at length converted into a brown mass, which falls to a coarse powder when touched.

The decomposition of woody fibre may take place in at least two conditions, namely, in the moist state subject to the free access of air, and under the surface of water, where the air obtains access only by being dissolved in the water; the products of the decomposition in the two conditions are different. When the access of air is limited, the decayed part does not acquire a brown colour, but becomes greyish-white, and a matter soluble in water is produced during the decay, the nature of which has not been minutely investigated.

If the wood is freely exposed to the air, it may suffer another kind of decomposition, known as the *dry-rot*, by which it is rendered brittle, and has its cohesion completely destroyed. If absolutely dry, the wood is not subject to this disease, but it is capable of absorbing sufficient moisture from the air to continue the decomposition when once commenced. In damp and ill-ventilated situations the dry-rot produces the most serious damage, causing the destruction, in a few years, of entire buildings and ships. But though favoured by a close and damp atmosphere, it may occur where the ventilation is perfect and the atmosphere in its usual state of humidity.

Dry-rot has been commonly attributed to the attacks of fungi; the disintegration being effected partly by the introduction of their filamentous spawn between the woody fibres, and partly by the moisture they are the means of conveying to the interior of the wood. In most cases of dry-rot, fungi are certainly present, but often not until some time after the commencement of the disease; in a few cases, however, fungi never make their appearance. Although, therefore, they may be intimately concerned in the rapid propagation of the disease when once commenced, they do not appear to be its original cause.

That the decay of wood, once dried, is in all cases produced by an external cause, is evident from the circumstance that if the surface of the dried wood is completely covered with a varnish impermeable to the air, decomposition is effectually prevented. Under the surface of water, if putrefying organic matter is absent, wood seems also to be able to resist decomposition for an indefinite time, and in dry air it may be kept for thousands of years without undergoing any sensible change.

In order to examine minutely into the means of guarding against the decay of wood, it will be found convenient to consider, in the first place, the characters and composition of wood in a healthy state, and afterwards, the chemical changes which take place in the processes of decay.

§ I. PROPERTIES AND COMPOSITION OF WOOD.

The function of the woody tissue of plants, in a physiological point of view, is to support the various deciduous organs for digestion, respiration, &c., being in this respect similar to that of bones in animals; to receive certain secretions, and to contain the sustenance necessary for the newly forming parts before a more direct communication is established between them and the soil. Though perfectly homogeneous to the naked eye, woody tissue is perceived, when examined by the microscope, to be composed of long, thin, transparent, tough, membranous tubes, which seem to be originally derived, like all other solid parts of plants, from a simple rounded cell. Woody tissue, though the chief and essential, is not the only constituent of wood or timber. Interspersed between the tubes which form the woody tissue, is the cellular tissue, which consists of cells or cavities closed on all sides, formed of a delicate and usually transparent membrane. Cellular tissue is more abundant in herbs than in trees, and decreases in proportion as the plant attains maturity. In *exogenous* trees it forms perpendicular plates radiating from the pith, as a centre, to the bark.

The cross section of a first year's stem of an exogen (to which class belong almost all trees producing woods employed for mechanical purposes in this country) presents, 1°, in the centre, pith, composed of cellular tissue; 2°, around the pith a layer composed principally of woody tissue; 3°, around the woody tissue a layer of bark (composed of several similar layers); and, lastly, 4°, from the pith to the exterior of the wood lines of cellular tissue, which are sections of the radiating plates above referred to, distinguished as medullary rays, medullary processes, or medullary plates. These are often imperceptible by the naked eye, but always present.

At the commencement of the second year's growth, a spontaneous separation of the layer of wood and the innermost layer of the bark or the *liber* takes place, and the intervening space becomes occupied with a viscid gelatinous liquid, known as the *cam-*

bium. In this liquid are deposited elongated cells or tubes, which form the woody tissue of another layer of wood immediately surrounding the first year's layer, and the principal part of the cellular tissue that connected the wood and the liber becomes arranged in perpendicular plates, forming continuations of the medullary rays of the first year. This second year's layer of wood is quite similar to that formed in the first year, to which it is firmly attached. The increase of wood goes on in this manner, circle around circle, or rather zone upon zone, each year; so that, with young trees, where the line of distinction between the several layers is easily perceived, the age of the tree may be estimated by the number of layers. Besides the external ring of wood, there is also formed yearly an internal ring of bark (liber), which at length becomes the external ring, those rings previously exterior to it having decayed as the stem increases in diameter. The name of the class of trees having this manner of growth, or exogens, has reference to the *external* augmentations of woody matter; unlike *endogens,* in which the wood is formed by successive augmentations from the interior.

The length of the medullary plates varies from a quarter of an inch or less, as in the sycamore and maple, to several inches, as in the oak. When viewed by the microscope with a low power, they present a granular appearance; but with a high power, a cellular structure similar to that of the pith is perceptible. The light and glossy appearance of polished vertical pieces of several kinds of wood, known among carpenters by the term silver-grain, or flower of wood, is produced by the exposure of the medullary plates.

The tubes which form the woody tissue vary in diameter from $\frac{1}{3000}$th to $\frac{1}{150}$th part of an inch. They taper acutely at each end, and do not appear to have any direct communication with each other; no pores are perceptible in their sides. They are very tough and usually cylindrical, but have sometimes been observed in a prismatic form.

The reason why the yearly increments of woody matter in exogens are defined (they being in juxtaposition and composed of a similar structure), is, that the woody tissue formed towards the close of the growing season is denser and more compact than that formed at the commencement. If, however, through an equable climate or any other cause, the tissue formed at the close of the season is quite similar to that formed at the commencement, no distinction between the yearly increments will be perceptible; in the wood of tropical countries, the absence of concentric circles is a very frequent occurrence.

In trees of less than eight or ten years old, there is usually no perceptible difference (excepting the lines of demarcation) between the several layers of woody tissue; but, after the lapse of ten or twelve years, the two or three interior layers become considerably hardened, and pass into the state of timber properly so called. The interior hardened layers are distinguished as the *duramen* and *heartwood*, and the softer exterior layers as the *alburnum* and *sapwood.*

The sides of the tubes of the woody tissue forming the alburnum are very thin, and hardly any solid matter is contained in the interior of the tubes, but merely sap ; the alburnum being the principal channel through which the sap is conveyed from the roots to the leaves. The alburnum is always lighter in colour than the duramen, and, having little solidity and power of adhesion, is readily susceptible of disintegration and decomposition ; on which account it is always separated from the heartwood when the timber is worked up. The superior hardness and durability of the heartwood is owing to the thickening of the sides of the tubes by the deposition of various solid matters ; as, the débris of disintegrated tubes and cells, resins, insoluble compounds of tannin with matters derived from the sap, colouring matters, &c., which impart peculiar characters to different species of woods : the sapwood has nearly the same appearance in all trees.

In some trees, as oak and teak, the conversion of sapwood into heartwood takes place rapidly; in others, as poplar and willow, very slowly, or not at all. The wood of the latter class of trees (which are technically called white-wooded) never acquires the durability of that of the former, and is unfit for any but temporary uses.

After the formation of heartwood is once commenced, the number of layers of sapwood usually continues the same at all stages of the growth of the tree ; consequently a layer of heartwood is produced annually. The heartwood itself is not of the same density throughout ; its interior layers gradually attain a maximum density, which is acquired by the other layers in yearly succession. After having remained for some time at the maximum density, the interior layers seem to lose their vitality, becoming lighter in colour, softer, weaker, and readily altered by the action of decomposing agents.*

Such is the structure and manner of growth of exogens. With reference to the growth of the other great class of trees, *endogens*, which includes palms, bamboos, grasses, &c., it will be sufficient for our purpose merely to mention that it is essentially different from that of exogens, the new woody matter being first developed towards the centre of the trunk; whence the name of the class. Endogens are not

* A difference of opinion has existed among botanists concerning the origin of woody tissue. According to the theory of Du Petit Thouars, as modified by Lindley and others, wood may be considered as the *roots* of the leaves and buds which are sent downwards through the cambium, and at length reach the extreme roots of the tree; by their close lateral adherence they form a layer which entirely surrounds the wood of the preceding year, and becomes itself a component part of the new wood. Consistently with this theory, the amount of wood is generally observed to be proportional to the amount of buds; and, if the leaves and branches which grow on one side of a tree are more vigorous than on the other (as may happen from exposure to more light and heat), the thickness of the layers of wood is greater on the side with vigorous leaves and branches than on the other. When the growth of the branches is equal on all sides, the thickness of the layers of wood is also usually equal all around. But leaves are not the only agents by which the woody tissue is developed, for many parts of plants and some whole orders (as cactaceæ) possess no leaves, and yet develope woody tissue. It has also been proved by Dr. Lankester, that trees from the stems of which the bark is removed at the spring of the year, will present new woody tissue between the bark and wood at the end of the year. Other circumstances may also be adduced in opposition to the theory which supposes buds to be the only agents concerned in the production of wood. The recent researches of Dr. Schleiden go far to prove that the original cells, which become elongated into tubes forming woody tissue, are developed in the same manner as the cells of the cellular tissue, that is, as excrescences proceeding from particles (cytoblasts) in the sides of anteriorly formed cells.

possessed of a well-defined cylindrical column of pith, nor of medullary rays: the densest part of their section is near the surface, instead of being near the centre, as in the heartwood of exogens.

Wood is unfit to be used for building in the state in which it is felled. The tissues, being then distended with sap, experience a contraction when the water in the sap evaporates; and, if the recently felled wood is placed in a confined situation, the humid nitrogenized matter in the sap rapidly decomposes, and induces the decomposition or decay of the wood.

To avoid these inconveniences, the wood, before being worked up, is carefully dried or "seasoned," by which it is reduced in bulk across the grain, and the nitrogenized matter of the sap is rendered less susceptible of decomposition. The ordinary process of seasoning wood consists in merely exposing it to a free current of air, the wood being either in the form of planks or logs, or in smaller pieces of about the sizes and forms to which they will afterwards be reduced. If the pieces are thin, twelve months' exposure in a dry situation with a free current of air will complete their desiccation to the extent required; but thick pieces often require several years. In general, the closer the grain, the longer is the time required; thus a large piece of oak is not thoroughly seasoned in less than eight or ten years. The exposure ought to be continued until the wood ceases to lose weight from evaporation, but this would require twice the period usually allowed for the process.

The seasoning of wood is said to be effected better and more rapidly by previously washing out or diluting the sap, which may be accomplished by exposing the wood for some weeks to running water, or by boiling the wood in water. A quantity of the soluble matter in the sap is brought to the surface when the wood is exposed to the action of steam, as in the operation for facilitating the bending of oak and other timbers for ship-building, &c.

A patent for an improved method of seasoning timber was obtained in 1825 by Mr. J. Langton of Lincolnshire, which consists in drying the wood in a vacuum, or in a highly rarefied atmosphere. The timbers are placed vertically in an air-tight cast-iron cylinder connected with an exhausting pump, and, when exhausted of its air, the cylinder is heated by means of a vapour-bath. The moisture given off from the wood is condensed in an air-tight refrigerator, so as to prevent its reabsorption.

The amount of contraction which takes place through desiccation is very different in different woods, being usually greatest in soft woods. In teak-wood the contraction is scarcely perceptible; in some soft woods it amounts to half an inch in the foot.

The entire proportion of water in green woods varies from 38 to 45 per cent., according to the species and age of the wood; but the whole of the water cannot be removed by drying in the air at common temperatures, however long the desiccation may be continued. Woods from the mulberry-tree, hazel-tree, and linden-tree, cut from branches of mean size at the close of autumn, decreased in weight in six months in the following proportions :—mulberry, 26 per cent.; hazel, 33 per cent.; and linden, 40 per cent. After being dried during twelve months, wood generally retains from one-fifth to one-fourth of its weight of water. A beam of oak-wood, kept for a century in a dry situation, was found by Count Rumford to lose 9 per cent. of its weight when dried at a high temperature. According to M. Karsten, oak shavings perfectly desiccated in the air lose 10·3 per cent. of water when heated to 212° ; but even at that temperature they retain a sensible quantity of water capable of being expelled at higher temperatures. The woods of the willow and birch, in a state of fine powder, and freed from sap by digestion in boiling water, retain 14·5 per cent. of water after desiccation in the air, for the expulsion of the whole of which the wood must be heated gradually to a temperature near 310° Fahr. (Prout.)

When wood, rendered perfectly dry by the aid of heat, is exposed at common temperatures to the atmosphere in its ordinary state of humidity, it reabsorbs a certain proportion of water, varying according to the compactness of the wood and to the quantity of deliquescent saline matters present. In a dry room without a fire, the quantity absorbed usually amounts to about 10 per cent. If covered with a resinous varnish, dry wood does not absorb atmospheric humidity.

In its ordinary state, wood is a conductor of electricity, from the presence of saline solutions ; when rendered perfectly dry by the aid of heat, it is a non-conductor, but its conducting power returns upon the absorption of moisture, which takes place on re-exposure to the air.

Although nearly all kinds of wood float on water, yet the density of the true woody fibre is considerably greater than that of water. The apparent lightness of wood is owing to the presence of a large quantity of air in the pores of the wood, which is not displaced by water at common atmospheric pressure without a very long digestion. But if a piece of wood is placed on water in the receiver of an air-pump, and a vacuum made, as the air in the pores of the wood is withdrawn, water enters the pores and the wood sinks.

According to Count Rumford, the specific gravity of the true woody fibre is much the same for all kinds of woods, varying only between 1·46, which is that of fir and maple, and 1·53, which is that of oak and beech. The specific gravity of the different kinds of woods in their ordinary state must therefore indicate their porosity, or the proportion of air within their pores. To take the specific gravity of wood in water for this purpose, the absorption of water by the wood should be prevented by applying to the surface of the wood a resinous varnish of the same density as water, which may be obtained by a mixture in certain proportions of wax and resin.* The first column in the following table exhibits the specific gravity of different woods as adopted by the *Annuaire du Bureau des Longitudes;* the second column contains the results obtained by M. Karmarsch.

* The specific gravity of wax is 0·967, and that of resin 1·079.

	I.	II.
Box	—	942
Plum-tree	—	872
Hawthorn	—	871
Beech	852	—
Ash	845	670
Yew	807	744
Elm	800	568
Birch	—	738
Apple	733	734
Pear	—	732
Yoke-elm	—	728
Orange-tree	705	—
Walnut-tree	—	660
Pine	657	763
Maple	—	645
Linden-tree	604	559
Cypress	598	—
Cedar	561	—
Horse-chestnut	—	551
Alder	—	538
White poplar	529	—
Common poplar	383	387
Cork	240	—

The same kind of wood varies considerably in density according to the soil on which the tree is grown, the climate, the age of the wood, and other circumstances. According to Rumford, the specific gravity of a piece of wood taken from the trunk of an oak in active growth is 961; that of billets of oak, cut and dried for a few years, is 883; that of a beam of oak, cut for at least six hundred years, was found to be 682; and that of the same wood when completely desiccated, 610.

Several exotic woods are considerably heavier than those kinds grown in Europe: the wood of the *guaiacum officinale,* for example, possesses the specific gravity 1·263; the specific gravity of ebony is 1213. Probably the heaviest of all woods is that known by the name of the iron-bark wood, brought from New South Wales, the density of which is 1426. Its strength, compared with English oak, was found by Mr. Holtzapffel to be as 1557 is to 1000. The lightest of true woods known in this country is the *Cortiça,* or *Anona palustris,* the density of which, according to Mr. Holtzapffel, is only 206. This wood resembles ash in colour; but is paler, finer, and softer.

From three to six per cent. of exsiccated wood is composed of solid matters derived from the evaporation of the sap, in which they were previously contained in a state of solution. These consist partly of saline matters, the proportion of which varies in different woods from two parts in a thousand to two per cent. But by far the principal part of the residue of the evaporation of the sap is a substance termed *vegetable albumen,* which closely resembles animal albumen (white of egg) both in properties and composition. It contains nitrogen, and, like animal albumen, is exceedingly prone to decomposition. The use of this substance in the living plant is to lubricate the sides of the various vessels, being the same as that of the mucous membrane of animals. It will be seen in another part of this article that the decay of woody fibre is generally an induced effect of contact with vegetable albumen in a state of decomposition. Different woods vary very considerably in the proportion of albumen which they contain.

It has been lately shewn, by M. Hartig, that a considerable quantity of *starch* is deposited in the interior of the vessels of the wood, which is capable of being extracted by mechanical means. The proportion of starch is said to be greatest in the winter season. To procure starch from this source, it is recommended to reduce to powder the dried shavings of green wood, and to rub the powder with a quantity of water. After standing for five or ten minutes, the ligneous powder may be separated by decanting the liquid, from which the starch is gradually deposited. Like all other varieties of starch, this substance is coloured intensely blue by iodine, and, when examined by the microscope, is perceived to be composed of spherical granules. The taste of its solution in warm water is slightly astringent.

By digesting the sawings of wood or the fibre of lint and cotton, successively in ether, alcohol, water, a diluted acid, and a diluted caustic alkali, so as to separate all the matters soluble in these liquids, without continuing the action of the acid and alkali so long as to alter essentially the constitution of the wood, there remains behind a white, spongy, pulverulent substance, which is the basis of the wood, or lignin, constituting from 95 to 97 per cent. of all kinds of desiccated wood.

Lignin is possessed of certain physical and chemical properties, which amply distinguish it from every other vegetable principle. These properties are always the same, if the lignin is prepared as above, however great the difference which may exist between the plants, or parts of plants, from which it is prepared. White unsized paper, digested in dilute hydrochloric acid to remove the earthy matters which it contains, and then washed with distilled water, affords a very pure form of lignin.

In a state of purity, lignin possesses the following properties:

It is white, tasteless, and inodorous, and presents, when examined by the microscope, a cellular or tubular structure. It is considerably heavier than water, but usually floats on that liquid in consequence of containing air imprisoned within its cells or tubes. It is insoluble in water, alcohol, ether, fixed and volatile oils, diluted alkalies and diluted acids. It dissolves in the most highly concentrated nitric acid, without producing the decomposition of the acid; and, if the solution is immediately diluted with water, it gives a white pulverulent precipitate, which is a neutral substance, highly combustible, insoluble in water, containing, according to Robiquet, the elements of nitric acid. Weaker nitric acid converts it into oxalic acid, suberic acid, and other products. When fused with a caustic alkali, lignin is converted into either ulmic acid or oxalic acid, according to the proportion of alkali and the temperature which is applied.

When lignin is mixed cautiously with concentrated sulphuric acid, so as to avoid elevation of temperature, it is converted partly into dextrin, a gummy substance which is produced by the action of dilute acids and other agents on starch. A portion of the sulphuric acid unites at the same time with some of the ligneous matter to form a compound which has received the names of *lignin-sulphuric acid*, and *vegeto-sulphuric acid*, which forms soluble salts with barytes and oxide of lead. When the above mixture of concentrated sulphuric acid and dextrin is diluted with water and boiled, the dextrin passes into the state of starch sugar.

To prepare starch sugar from this source, six parts of clean hempen or linen cloth, divided into small pieces, is intimately mixed with eight and a half parts of concentrated sulphuric acid, added in very small quantities. In the course of half an hour, when the cloth has become converted into a brown viscous mass, entirely soluble in cold water, sufficient water is added to dissolve the mass, and the mixture is boiled for eight or ten hours, fresh water being added from time to time to replace that which is expelled by evaporation. The saccharification is then complete, and the free sulphuric in the solution is separated by the addition of an excess of chalk, which becomes converted into the insoluble sulphate of lime. The filtered liquid leaves a residue of starch sugar on evaporation. According to M. Braconnot, twenty parts of lignin afford about twenty-three parts of sugar.

Lignin resembles starch not only in being convertible into dextrin and sugar by the action of acids, but also in being converted into dextrin (or an analogous substance) by mere torrefaction.

The identity in the nature of the ligneous matter derived from all kinds of trees is evidenced not only by its chemical properties, but by its analysis. The composition of washed and dried wood from the oak and beech was found by MM. Thénard and Gay-Lussac to be the following:

Carbon				51·45
Hydrogen	5·82	} = water		48·55
Oxygen	42·73			
				100·00

Analyses of washed and dried wood from the box and willow afforded Dr. Prout similar results.

From these and other analyses the formula $C_{36} H_{22} O_{22}$ has been assigned to pure lignin.

According to the more recent researches of M. Payen, the lignin of wood is not a homogeneous substance, but a mixture of two principles, which he has succeeded in separating. One of these is the primitive woody tissue, which has the same composition as starch ($C_{12} H_{10} O_{10}$), and is named *cellulose* by M. Payen. The other principle is the true ligneous matter, and is contained in the interior of the cells in a proportion varying according to the nature of the wood and the age of the tree. It forms the principal part of the solid matter which gives durability and hardness to the heartwood. The separation of the true lignin (matière incrustante) from the cellulose was effected by M. Payen by the agency of highly concentrated nitric acid, which dissolves lignin, but leaves the cellulose without alteration. The remaining cellulose is soluble in concentrated sulphuric acid without being blackened. The true lignin is considered by M. Payen to contain more hydrogen than is required to form water with the oxygen present; that from hemp, lint, straw, and linen cloth corresponded in composition with the formula $C_{35} H_{24} O_{20}$.

When wood is subjected to destructive distillation in close vessels, a great variety of volatile products is disengaged, comprising water, acetic acid, a black mass known as wood-tar (containing several peculiar combustible liquids and solids), several oily and spirituous substances, carbonic oxide gas, and gaseous compounds of carbon and hydrogen. The higher the temperature at which the distillation is conducted, the larger is the proportion of gaseous matters among the products. A residue of charcoal of the same shape as the original pieces of wood, but less in size, is found in the retort, the weight of which varies from 16 to 28 per cent. of the weight of the wood.

The *flame* of burning wood proceeds from the combustion of the same kind of gaseous matters as are given off when the wood is subjected to destructive distillation in close vessels. As the proportion of these products is partly dependent on the temperature at which the distillation is conducted, it follows that, to obtain the largest possible flame, the wood should be dry, in order to avoid loss of heat by the evaporation of the water, and in small pieces which may be quickly heated to their centre and applied to the fire in small quantities at a time. If the temperature necessary for active combustion is maintained, and sufficient air has access, the combustion of the wood is complete; the only residue being a small quantity of white ash derived from the saline and earthy matters formerly contained in the sap. The carbon of the wood in this case is entirely converted into carbonic acid, and the hydrogen into water, by combining with the oxygen of the air and of the wood. But it is difficult to unite at all times the conditions necessary for perfect combustion, namely, a high temperature and sufficient air; the combustion or oxidation of the volatile products is hence often incompletely effected, and *smoke** (which consists chiefly of solid particles of a carbonaceous substance) is produced. Compact

* The soot which is deposited on a cold body from the smoke of burning wood has been analyzed by M. Braconnot with the following results:

A nitrogenized carbonaceous matter insoluble in alcohol	20·00
Ulmin	30·20
An acrid and bitter principle	·50
Water	12·50
Carbonaceous matter insoluble in alkalies	3·85
Acetate of potash	4·10
Acetate of ammonia	·20
Acetate of lime	5·65
Acetate of magnesia	·53
Acetate of iron	a trace.
Chloride of potassium	·36
Sulphate of lime	5·00
Ferruginous phosphate of lime	1·50
Carbonate of lime	14·66
Silica	·95
	100·00

Lamp-black differs from common soot in being more completely carbonized. According to the analysis of M. Braconnot, the composition of lamp-black from wood is the following:

Carbon	79·1
Water	8·0
Resin	5·3
Bitumen	1·7
Ulmin	·5
Sulphate of potash	·4
Sulphate of ammonia	3·3
Sulphate of lime	·8
Chloride of potassium	a trace
Ferruginous phosphate of lime	·3
Quartzy sand	·6
	100·0

woods burn only at the surface; the volatile combustible products which produce flame are quickly disengaged, and a mass of charcoal remains, which burns away slowly without the production of flame, or at least of the yellow flame which is perceived at the commencement of the combustion. Light porous woods, which freely admit air to their interior, burn more rapidly than compact woods, and afford a yellow flame almost the whole time of their combustion, leaving a very small residue of charcoal.

With a view of determining the heating power of different kinds of wood in a state of combustion, a set of experiments was performed by MM. Peterson and Schödler to ascertain the quantity of oxygen required for the combustion of a given weight of the different woods. If the woods are equally dry, the amount of heat disengaged by the combustion is very nearly proportional to the quantity of oxygen which unites with the combustible. The results obtained by MM. Peterson and Schödler are the following:

Names of Trees.	*Oxygen required to burn* 100 *parts of each.*
Tilia Europea, lime .	140·523
Ulmus suberosa, elm .	139·408
Pinus abies, fir . .	138·377
Pinus larix, larch .	138·082
Æsculus hippocastanum, horse-chestnut .	138·002
Buxus sempervirens, box	137·315
Acer campestris, maple	136·960
Pinus sylvestris, Scotch fir	136·931
Pinus picea, pitch pine	136·886
Populus nigra, black poplar	136·628
Pyrus communis, pear-tree	135·881
Juglans regia, walnut	135·690

Names of Trees.	*Oxygen required to burn* 100 *parts of each.*
Betula alnus, alder .	133·959
Salix fragilis, willow .	133·951
Quercus robur, oak .	133·472
Pyrus malus, apple-tree	133·340
Fraxinus excelsior, ash	133·251
Betula alba, birch .	133·229
Prunus cerasus, cherry-tree	133·139
Robinea pseudacacia, acacia	132·543
Fagus sylvatica, white beech	132·312
Prunus domestica, plum	132·088
Fagus sylvatica, red beech	130·834
Diospyros ebenum, ebony	128·178

Heat is dissipated from a body in a state of ordinary combustion in two ways entirely distinct from each other: 1°, by the ascending current of air and gases from the combustible; and, 2°, by radiation in all directions, both from the surface of the burning body and from the flame. M. Péclet has ascertained by experiment that the radiating power of burning wood is not the same for different kinds of wood; and that for the same kind it is greater when the wood is in large than in small masses, because the radiating power of the charred surface of the wood is much greater than the radiating power of flame. If the wood is in very small pieces, the amount of heat radiated is very nearly the same for all kinds of wood, and is equal to one-third of that dissipated by the draught of air, or to one-fourth of the entire amount produced.

The quantity of white ash which remains after all the combustible matter in wood is completely consumed is very small, and varies considerably from different woods. The following table shews the quantity which remains after the combustion and incineration of 1000 parts of several different kinds of wood.

Oak (faggots)	22	Spindle	4
Ebony	16	Ash	4
Mahogany	16	Filbert	4
Aspen	6	Box	3·6
Pine	4	Poplar	2
Birch	4	Cork	2
Fir	4	Maple	2
Barked oak	4		

These ashes are composed of alkaline salts soluble

in water, and earthy matters insoluble in water. The soluble alkaline salts consist of carbonates and sulphates of potash and soda, chlorides of potassium and sodium, and a little silicate of potash. The insoluble earthy matters are composed of lime or carbonate of lime, according to the temperature at which the incineration is performed, phosphate of lime, phosphate of magnesia, phosphate of iron, oxide of manganese, and silica. The proportions of ash given in the preceding table do not represent the actual quantities of saline matter in the different woods; as the greater part of the potash, soda, and lime exists in the woods in combination with vegetable acids, as tartaric, oxalic, acetic, &c., which become destroyed by the combustion. One of the products of the decomposition of the vegetable acids is carbonic acid, part of which combines with the bases formerly united to the vegetable acids. This is the only source of the carbonic acid in the carbonates of potash and soda found in the ash; as the wood before calcination does not contain a trace of an alkaline carbonate. The results given in the preceding table are those obtained when the ash is calcined at a full white heat, by which the carbonate of lime becomes converted into quick-lime. When thus calcined, the ash weighs nearly one-third less than it does if the residue of the combustion is simply incinerated at a red heat, until the charcoal is completely consumed. The relative proportion of ash obtained from the young and old wood of a few different kinds of trees is shewn in the following table of the results of some experiments by M. Karsten:

From young oak-wood	15	parts of ash in 1000 of wood.
„ old „	11	„
„ young red beech-wood (rothbuchenholz)	37·5	
„ old „	40	
„ young white beech-wood (weissbuchenholz)	32	
„ old „	35	
„ young alder-wood	35	
„ old „	40	
„ young birch-wood	25	
„ old „	30	
„ young red pine (fichtenholz)	15	
„ old „	15	
„ young white pine (tannenholz)	22·5	„
„ old „	25	
„ young pitch pine (kiefernholz)	12	„
„ old „	15	„

The difference in the composition of the ash derived from the wood of different kinds of trees is not very considerable. The portion soluble in water comprises from one-eighth to one-fourth of the entire ash. Analyses of the insoluble portions of the ashes of oak-wood and beech-wood afforded the following results (M. Berthier):

	From oak-wood.	*From beech-wood.*
Lime, as carbonate and phosphate	53·2	42·6
Magnesia	—	7·0
Oxide of iron	—	1·5
Oxide of manganese	9·0	4·5
Carbonic acid (as carbonate of lime)	24·4	32·9
Phosphoric acid (as phosphate of lime)	7·0	5·7
Silica	6·4	5·8
	100·0	100·0

The soluble portions of the same ashes were found to be composed of

	From oak-wood.	From beech-wood.
Potash and soda	65·2	64·1
Carbonic acid	24·3	22·4
Sulphuric acid	10·0	7·3
Muriatic acid	·5	5·2
Silica	—	1·0
	100·0	100·0

§ II. NATURE AND CAUSES OF THE DECAY OF WOOD.

By referring to the observations contained in the preceding section on the structure and composition of wood in a healthy state, it will be perceived that, according to the analyses of MM. Gay-Lussac and Thénard, and Dr. Prout, pure woody fibre may be considered to contain carbon united with the elements of water; but that, besides the pure fibre, wood contains, in its ordinary state, from three to five per cent. of soluble matter, of which nitrogen forms an essential constituent.

It appears from the experiments of De Saussure that *the decay of woody fibre is essentially a process of oxidation.* On exposing moist wood to oxygen gas, it was found that, for every volume of oxygen absorbed by the wood, one volume of carbonic acid gas was disengaged. As carbonic acid gas contains its own volume of oxygen gas, the result might seem to be merely a separation of a portion of the carbon of the wood by a process of oxidation or combustion at low temperatures. Such cases of slow oxidation have been distinguished by Liebig by the name of *eremacausis*, which is compounded from ἠρέμα, by degrees, and καῦσις, burning.

But the examination of the products of the action of oxygen on dry wood, and the analysis of the residuary mould or humus, shew that wood loses by the process of slow combustion or eremacausis, besides carbon, a certain amount of its hydrogen and oxygen in the proportion to form water. On exposing 240 parts of dry saw-dust of oak-wood to the action of oxygen gas, De Saussure found that the oxygen became converted into the same volume of carbonic acid gas, containing 3 parts by weight of carbon. But the wood diminished in weight by 15 parts; 12 other parts were therefore separated in the form of oxygen and hydrogen.

For the following considerations on the nature of the decay of woody fibre, I am indebted to Professor Liebig's valuable work entitled "Organic Chemistry applied to Agriculture and Physiology." In an agricultural point of view this process of decay is of great importance, as being the means of furnishing the soil with humus; which, by the action of the air, becomes a means of supplying the roots of plants with carbonic acid.

Notwithstanding the separation of carbon from wood during the process of decay, in the form of carbonic acid, if the composition of the decaying wood is examined at different stages in the process, it is found that the relative proportion of carbon in the different products augments as the decay advances. The weight of the hydrogen and oxygen therefore, which are given off simultaneously with the carbon, is greater than the weight of the separated carbon. According to Gay-Lussac and Thénard, 100 parts of oak-wood dried at 212°, from which all soluble matters had been separated by means of water and alcohol, contained 51·45 parts of carbon ; the remainder being hydrogen and oxygen in the same proportion as in water. A specimen of mouldered oak-wood, taken from the interior of the trunk, of a chocolate-brown colour and retaining the structure of wood, was found by M. Meyer to contain 53·36 per cent. of carbon, and 46·44 per cent. of hydrogen and oxygen, in the same proportions as exist in water. Another specimen of mouldered oak-wood, in a more advanced state of decay, of a light brown colour and easily reducible to fine powder, was found by Dr. Will to contain 56·2 per cent. of carbon, and 43·8 of hydrogen and oxygen in the usual proportions. The composition of these substances in equivalents is represented by the following formulæ, with which the percentage composition, as obtained by analysis, closely agrees.

Fresh oak-wood, by Gay-Lussac and Thénard	C_{36}	H_{22}	O_{22}
Humus from oak, by M. Meyer	C_{35}	H_{20}	O_{20}
Humus from oak, by Dr. Will	C_{34}	H_{18}	O_{18}

It appears from these data that, for each equivalent of carbon separated from the wood, there are also separated two equivalents of oxygen and two equivalents of hydrogen. The process is merely a case of slow combustion through the oxygen of the air, but it remains to be decided whether the carbon or the hydrogen of the wood unites with the oxygen absorbed from the air. One of these elements is doubtless oxidized by the air, the other unites more intimately than before with the oxygen of the wood. If, as some suppose, the hydrogen and oxygen of the wood already exist in the state of water, the wood being a hydrate of carbon, there can be no doubt that the carbon is oxidized immediately by the air. But the characters of woody fibre favour the idea that its hydrogen and oxygen do not exist in the form of water; for, were that the case, dry starch, sugar, and gum must likewise be considered as hydrates of carbon. And, if the hydrogen does not exist in woody fibre in the form of water, the direct oxidation of the carbon cannot be considered as at all probable without rejecting all the facts established by experiment regarding the process of combustion at low temperatures. (Liebig, Agricultural Chemistry, p. 295, 3rd edit.)

If such were the case, it would be a combustion in which the carbon of the burning body constantly augmented, instead of being diminished. It may therefore be concluded that it is the hydrogen which is oxidized at the expense of the oxygen of the air, while the carbonic acid is wholly formed from the elements of the wood.

It is uncertain to what extent this decomposition of the wood may proceed under the ordinary influence of moisture and air, favoured with warmth and light, but certainly not so far as to the entire separation of the hydrogen and oxygen; for, with the increased proportion of carbon in the residuary humus, its affinity for the remaining hydrogen increases, until at length it equals the affinity of the oxygen for the same element.

Such are the changes which wood suffers when exposed in a humid state to an unlimited quantity of air. But when the air is entirely or partially excluded, water still having access, the order of decomposition is considerably modified. Carbonic acid is then evolved in the same manner as when air is freely admitted, but the hydrogen of the wood remains behind. Two analyses of the white decayed wood, obtained from the interior of the trunk of a dead oak which had been in contact with water, yielded, after having been dried, at 212°,

Carbon	47·11	48·14
Hydrogen	6·31	6·06
Oxygen	45·31	44·43
Ashes	1·27	1·37
	100·00	100·00

The mean of these numbers corresponds very nearly to the formula $C_{33}\ H_{27}\ O_{24}$, which gives, by calculation, carbon 47·9, hydrogen 6·1, oxygen 46. By comparing the formula with that assigned to woody fibre by Gay-Lussac and Thénard, it is seen that the elements of water become united with the wood, and a portion of the carbon of the wood is separated in the form of carbonic acid. The oxygen of the carbonic acid is derived partly from the wood and partly from other sources, particularly from free oxygen contained in solution in the water, and derived from the atmosphere. It is supposed by Liebig that in this, as in all other cases of putrefactive decomposition, the oxygen of the water itself assists in the formation of the carbonic acid.

The above formula for mouldered oak is obtained by adding to the formula for pure woody fibre the elements of five equivalents of water, and three equivalents of oxygen, and subtracting therefrom three equivalents of carbonic acid.

	C	H	O
Woody fibre	C_{36}	H_{22}	O_{22}
5 eq. water		H_5	O_5
3 eq. oxygen			O_3
	C_{36}	H_{27}	O_{30}
Deduct 3 eq. carb. acid . . .	C_3		O_6
Mouldered oak	C_{33}	H_{27}	O_{24}

But the composition of mouldered wood varies according to the facility with which oxygen has access. A specimen of white mouldered beech-wood afforded, on analysis, carbon 47·67, hydrogen 5·67, and oxygen 46·68, numbers which correspond to the formula $C_{33} H_{25} O_{24}$ (Liebig).

The decomposition which wood suffers in marshy soils, and under water in contact with decaying vegetable matter, the access of free oxygen being wholly prevented, is a case of putrefaction or transformation of the constituents of the wood into other forms of matter, differing from the preceding cases in the circumstance that the decomposition is not effected by the separation of one of the elements of the wood through the affinity of an external substance. In the cases already considered, the decomposition is partly effected through the affinity of oxygen for the hydrogen and carbon of the wood. When the wood is imbedded in marshy soils, the access of free oxygen is prevented by the wood being surrounded by attenuated decomposing vegetable matter, which has a more powerful affinity for oxygen than the denser wood.

In such cases of transformation, the carbon of the wood is shared between the hydrogen and oxygen of the wood ; the carbon forming with hydrogen light carburetted hydrogen or marsh gas (CH_2), and with oxygen, carbonic acid. A portion of the hydrogen and oxygen of the wood may probably unite to form water at the same time. The mouldy residue always contains a much larger relative proportion of carbon than previously existed in the wood.

In assigning a cause for these transformations, it is not sufficient to ascribe them to the action of air and water. It is known that in dry air woody fibre may be preserved without decomposition for thousands of years ; and under water, in certain conditions, it appears to be equally durable.

The general condition for the production of such decompositions is *contact with a body already undergoing a similar change.* When fresh wood is placed in contact with decaying wood, or other decaying vegetable matter, the latter acts the part of a ferment, and causes the transformation of the elements of the fresh wood, from the same cause as yeast causes a transformation of the elements of sugar. The result differs according to the composition of the substance to be decomposed, and the presence or absence of free oxygen ; but the cause is the same, and the decay in the case of dry-rot may be considered as a process of fermentation and oxidation combined.

In this respect it closely resembles acetification or the process of fermentation and oxidation, by which alcohol is converted into acetic acid. By absorbing oxygen, the alcohol becomes entirely converted into water and acetic acid. But pure alcohol, whether strong or diluted with water, may be exposed to free oxygen without the formation of the smallest particle of acetic acid, or alteration of any kind; as also may pure woody fibre without formation of humus. One of the conditions necessary to the conversion of alcohol into acetic acid is contact with a nitrogenized body, as yeast, in a state of slow oxidation or putrefaction; a condition precisely similar to that under which the decomposition of wood is originated. The transformation of the elements of alcohol, or woody fibre, is considered by M. Liebig as a reflex action of the transformation of the contiguous decomposing body consistently with the physical law proposed in another application, by Laplace and Berthollet, that "a molecule set in motion by any power can impart its own motion to another molecule with which it may be in contact."

The albuminous matter which wood contains distributed over the cellular tissue is intimately connected with the decomposition of the wood. It is a nitrogenized substance, and identical in composition with animal albumen; like which body, it is putrescible in a high degree, and therefore an element of fermentation and putrefaction. It is particularly adapted for food for insects, which are often found in the interior of the cells, penetrating the wood in all directions. The disintegration of the fibres, thus occasioned by insects, also greatly accelerates the chemical action going on at the same time, from the increased facility for the introduction of air and water to the interior of the wood. The products of the spontaneous decomposition of vegetable albumen are the same as those produced from animal albumen; namely, carbonate of ammonia, nitrate of ammonia, carburetted hydrogen, and water. The influence of ammoniacal salts in favouring the growth of fungi accounts for the appearance of the latter among the earliest signs of decay of wood.*

The spontaneous decomposition of the vegetable albumen, acting as a ferment, is the primary cause of the decomposition of the wood; and the presence of sugar, starch, and other matters capable of being easily transformed by a ferment, considerably hastens the decomposition. Hence it is found that those woods which contain the smallest quantity of albumen and amylaceous matters are the most durable. The wood of a tree of the acacia tribe, which has been largely employed of late in France and America for purposes in which the wood is subject to more than ordinary exposure, contains merely a trace of albumen, and hence resists decomposition in situations where all other woods enter into a state of decay. Placed in the same circumstances in which oak-wood decays in one or two years, the wood of the acacia is said to remain perfect for fifteen or twenty years (M. Payen †).

Before a considerable quantity of wood is appropriated to building purposes, an experiment should be performed to ascertain, by chemical analysis, the proportion of albumen which it contains. A method recommended by M. Payen consists in digesting the wood with the aid of heat in a dilute solution of caustic soda, which has little or no action on the woody fibre, but dissolves the albumen; and the

* The growth of mushrooms is found to be remarkably accelerated by watering them with a solution of sulphate of ammonia.

† *Cours de Chimie Organique*, 1843.

quantity of the latter thus separated may be estimated by washing, drying, and weighing the wood.

As the presence of starch and gum in the wood would prevent such a process as the preceding from affording anything more than an approximation to the proportion of albumen, a more advantageous method would be to determine the absolute amount of nitrogen in the wood by the simple and easy process devised by MM. Will and Varrentrapp for the ultimate analysis of organic bodies. From the proportion of nitrogen thus obtained, that of the albumen in the wood may be calculated; but such a calculation is unnecessary, as the relative amount of nitrogen in different specimens of wood would of course indicate (according to the principles before developed) their relative disposition to decay. An account of the manner of conducting the process referred to may be found in the "Elements of Chemical Analysis," page 267

§ III. PRESERVATIVE MATERIALS.

If the decay of wood is, in the first place, an induced effect of the contact of decomposing albumen, a means of preserving the wood is naturally suggested in the removal of the albumen; or else, in so modifying it, by causing it to combine with other substances, that it shall no longer possess the property of decomposing spontaneously.

The solubility of albumen in cold and tepid water affords a simple means of withdrawing from the wood this element of decomposition. Unless the wood is in very thin pieces, however, the removal of the albumen by the process of washing in water is extremely slow. To test the efficacy of merely washing in water, equal weights of washed and unwashed wood, equally dry, were moistened with the same quantity of water, and the amount evaporated was replaced in each quantity equally. In the course of a few weeks the unwashed wood was always found to be covered with a thick mould, while none appeared on the washed wood for six months. At the expiration of that period the unwashed wood was found to have sensibly diminished in weight, while the weight of the washed wood remained unaltered (Dr. Boucherie). As the decay of wood advances, the proportion of soluble matter decreases from five or six to less than one per cent.

But, as the removal of the albumen seems to diminish the adhesion of the fibres and the tenacity of the wood, a better method of preserving wood is to cause the albumen to enter into combination with another substance, to form a compound which is insoluble in water, and not susceptible of spontaneous decomposition. This is the mode of action of all the antiseptic substances which have been of late applied to wood, either in aqueous solution or in the form of vapour, as effectual preventives of decay.

Corrosive sublimate, or chloride of mercury, is one of the most efficient of these antiseptic applications It was proposed by Mr. Kyan* as a preventive of

* The use of corrosive sublimate was previously suggested to the Admiralty and Navy Board by Sir H. Davy; but, this substance being slightly volatile at common temperatures, it was considered that the at-

dry-rot, under the idea of its acting as a poison to the fungi and insects which were the supposed cause of the disease. But this explanation of the action of corrosive sublimate is no longer tenable, as it is now generally admitted that the fungi and insects are not to be considered the origin, but the result, of dry-rot. It has been suggested that its action depends on the formation of a compound of lignin, or pure woody fibre, with corrosive sublimate, which resists decomposition in circumstances where pure lignin is liable to decay ; but pure lignin possesses no tendency to combine with corrosive sublimate. The action of this substance is in reality confined to the albumen, with which it unites to form an insoluble compound not susceptible of spontaneous decomposition, and, therefore, incapable of exciting fermentation.* Vegetable and animal matters, the most prone to decom-

mosphere surrounding the prepared timbers might become vitiated, and the proposal was not at that time carried into execution.

* A compound quite similar to, if not identical with, that referred to in the text, is thrown down as a white precipitate when an aqueous solution of corrosive sublimate is mixed with a solution of white of egg in water. It is not, as was formerly supposed, a compound of calomel with albumen, nor is it a direct combination of corrosive sublimate with albumen ; but a mixture of two substances, namely, albuminate of mercury, and muriate of albumen. By washing the precipitate with pure water, the muriate of albumen may be separated from the precipitate, and albuminate of mercury remains behind, mixed with a small trace of calomel formed by the reaction of a little phosphorus in the white of egg on the corrosive sublimate. If the solution of corrosive sublimate is mixed in large excess with the white of egg, the muriate of albumen which is formed remains in solution.

The albuminate of mercury is quite insoluble in pure water, and so inert that the poisonous effects of corrosive sublimate may be counteracted by swallowing the whites of several eggs immediately after the exhibition of the poison. It is soluble in water containing chloride of sodium (and therefore in sea-water), chloride of potassium, and muriate of ammonia.

position, are completely deprived of their property of putrefying or fermenting by the contact of corrosive sublimate. It is on this account advantageously employed as a means of preserving animal and vegetable specimens. Its expensiveness in this country is a great obstacle to its extensive employment, but few antiseptic applications are more effectual. In Mr. Kyan's process the wood to be impregnated is sawed up into blocks or planks, and soaked for seven or eight days in a solution containing one pound of corrosive sublimate to five gallons of water. The impregnation is sometimes effected in an open tank, and sometimes in an air-tight vessel from which the air is first exhansted by a pump as far as possible ; and the solution is then pressed into the pores of the wood under a force of about a hundred pounds to the square inch.

To test the efficacy of Mr. Kyan's process, protected and unprotected pieces of timber were placed in a trench in the Royal Arsenal at Woolwich in contact with putrefying vegetable matter, and with pieces of wood affected with dry-rot ; and the trench was covered with horse-dung to increase the temperature and accelerate the decomposition. At the expiration of five years the protected wood was found to be unaltered, while the same kind of wood, unprotected, became considerably affected before the end of the first year.

The action of almost all beneficial materials for impregnating wood may be considered of the same nature as that of corrosive sublimate.

The most ancient means of preserving wood consists in the application of an external resinous or oleaginous covering, by which air and water are effectually excluded. If the wood is dry and in a sound state before the covering is applied, perfect protection might be thus afforded, provided the wood is not exposed to abrasion. It is essential that the wood be made thoroughly dry previous to the application of a protective varnish, else its decay is hastened by the impediment which the varnish offers to the evaporation of the moisture.

The more effectual method of impregnating the wood throughout its mass with a chemical preservative agent was not practised to any great extent until the last century. The principal substances which have been proposed for that purpose are the following. (See Mr. John Knowles's "Inquiry into the means which have been taken to preserve the British Navy, particularly from Dry-rot," 1821.)

Tar.
Sulphate of copper.
Sulphate of iron.
Sulphate of zinc.
Sulphate of lime.
Sulphate of magnesia.
Sulphate of barytes.
Sulphate of soda.
Alum.
Carbonate of soda.
Carbonate of potash.
Carbonate of barytes.
Sulphuric acid.
Acid of tar (pyroligneous acid).
Common salt.
Vegetable oils.
Animal oils.
Coal oil (naphtha).
Resins.
Quick-lime.
Glue.
Corrosive sublimate.
Nitrate of potash.
Arsenical pyrites water (containtaining arsenious acid).
Peat-moss (containing tannin).
Creosote and eupion.
Crude acetate or pyrolignite of iron.
Peroxide of tin.
Oxide of copper.
Nitrate of copper.
Acetate of copper.
Solution of bitumen in oil of turpentine.
Yellow chromate of potash.
Refuse lime-water of gas-works.
Caoutchouc dissolved in naphtha.
Drying oil.
Bees'-wax dissolved in turpentine.
Chloride of zinc.

Sulphate of copper (blue vitriol), sulphate of iron (green vitriol), and sulphate of zinc have been employed for a considerable time as preservatives against dry-rot. An objection to sulphate of iron, and especially the persulphate, has been suggested by M. Bréant, in its property of being decomposed into the insoluble subsulphate of iron and free sulphuric acid by the woody fibre, which combines with the subsulphate, while the free sulphuric acid exercises a corrosive action on the timber, and often causes it to become almost pulverulent. These inconveniences may be obviated by first injecting some oily material into the pores of the wood. In the process for preserving animal and vegetable matters from decay, patented by Mr. Margary in 1837, the wood, previously dried, is soaked in a solution of one pound of sulphate of copper in five gallons of water; and is allowed to remain in the liquid two days for every inch of its thickness. Instead of the above solution, another is sometimes made use of, composed of a pound of acetate of copper dissolved in fourteen quarts of water with two quarts of crude pyroligneous acid. Like chloride of mercury, sulphate of copper acts by forming an insoluble and stable combination with albumen.

Contact with alkalies and alkaline earths greatly accelerates the decay of wood, for these substances enable woody fibre and several other organic matters to absorb oxygen, which do not possess the power of themselves. Thus alcohol, which does not, if pure, absorb oxygen from the air at common temperatures, when mixed with potash, absorbs it with avidity, becoming converted into acetic acid, formic acid, and other oxidated products. Several vegetable colouring matters, gallic acid, tannin, and other substances, are affected by alkalies in a similar manner.

An experiment on the durability of timber seasoned with lime was made some years ago on a part of the frame-work and some of the timbers of the Amethyst frigate. At the expiration of ten years the prepared timbers were found to be in a worse condition than the unprepared.

Alum effectually counteracts the decomposition of the albuminous matter of the wood; but it cannot be employed as a preservative material, from its decomposition, under the influence of the woody fibre, into the insoluble subsulphate of alumina, which attaches itself to the fibre, and free sulphuric acid, which exerts a corrosive action on the wood. The soluble subsulphate of alumina (basic alum) would be free from this objection to common alum, and, if made without excess of alkali, might probably be found an efficient application.

The antiseptic property of common salt is not without an application in the preservation of wood. The durability of the beams and other timber-work in salt-mines is attributable to the action of the salt in restraining decomposition. If kept in a tolerably dry atmosphere, wood impregnated with a solution of common salt resists decomposing agents for a considerable time, and it has been observed that ships employed in the salt-trade are more durable than most others built of the same kind of timber. For a first cargo for ships built along the shores of the Baltic the preference is generally given to salt.

But the deliquescent nature of this substance, in the state it is commonly met with, prevents its employment as a preservative for wood intended for general purposes. For buildings, however, in which the temperature is usually high, wood thus prepared would be found durable and economical. Pure salt has very little deliquescent property; that possessed by the salt met with in commerce is chiefly due to traces of chloride of calcium and chloride of magnesium.

The durability of ships employed in the salt-trade has been referred to the thorough desiccation of the timbers by the hygroscopic property of common salt. More than ordinary durability is also ascribed to ships employed in conveying quick-lime; an effect of lime which is hardly referable to any other mode of action, for the impregnation of the wood with lime-water would only facilitate its decay.

The saturation of wood with vegetable and animal oils, with a view to its preservation, has been practised in America to a considerable extent. This application of oil appears to have been known to the ancients, and was recommended by Dr. Hales in 1756. The imbibition of the oil by the wood is extremely slow, but the protection thus afforded is very considerable. To facilitate the impregnation, it has been proposed first to expel the air and moisture in the wood by the application of heat; and, as a temperature approaching 600° Fahr. may be attained in an oil-bath, the oil has been made the medium of drying and expelling the air as well as of impregnating the timber; but the wood which had undergone this process was found to have diminished in tenacity, and its fibres were easily separated from each other. (See Knowles's "Inquiry," before referred to.)

In 1811 a proposal was made by Mr. Lukin to impregnate wood with the vapour produced from fixed oils, and extensive works were erected for preparing wood for Government purposes by such a process. But most of the timbers submitted to the vapour became cracked, and rendered quite unfit for the construction of ships. The building in which the impregnation was effected, the length of which was thirty-two feet, and the breadth twelve feet, at last exploded; but the trial was quite adequate to prove the insufficiency of the process.

From the general nature of the action of arsenious acid on animal and vegetable substances, its efficacy as a preservative material for wood may be assumed. Mundic water containing arsenic, produced by the oxidation, through the air, of arsenical iron pyrites in contact with water, was proposed for this purpose by Mr. Lukin in 1812; but the use of this material was abandoned from its injurious effects on the workmen, the death of two persons being produced through some preliminary experiments to determine the value of the process.

The durability of wood is greatly increased by being impregnated with tannin, which acts on the albuminous matter in the same manner as corrosive sublimate. The preservation, for several ages, of large branches and trunks of trees imbedded in peat is wholly referable to this action of the tannin and analogous substances contained in the peat. With a view of producing in fresh oak-wood the same change which it experiences in bogs, it has been proposed to keep the wood surrounded for some time with peat-moss; but the experiments undertaken to test the efficiency of the process were failures, from the difficulty of carrying the impregnation to any extent. The wood which is taken from bogs, however, when exposed to the weather, becomes weak in the fibre, splits, and is soon impaired in quality (Mr. Knowles).

The remarkable antiseptic property of creosote has suggested the application of this substance as a preservative for wood. Creosote is an unctuous liquid found among the products of the distillation of wood, and is contained in the tar of some kinds of wood to the amount of one-fifth or one-fourth of the weight of the tar. In an impure state creosote may be obtained by merely subjecting wood-tar to redistillation and rejecting the first products, but for its preparation in a pure state, a more complicated process is necessary, an account of which, together with the properties of this singular substance, will be found in the article on the products of the distillation of wood.

The efficacy of tar as an external application to wood may be principally referred to the action of the creosote and eupion which it contains on the albuminous matter in the wood, in the same manner as corrosive sublimate. With the view of effecting a deeper impregnation, it has been proposed to steep the wood in boiling tar; but exposure to a boiling liquid for a short time has always the effect of diminishing the tenacity of wood. On comparing the strength of two pieces of timber, one having been boiled in tar, and the other in its ordinary state, but quite similar in other respects, the strength of the boiled timber was found to be one-seventh less than that of the unboiled. The wood, however, is rendered better capable of resisting decomposition, and suffers an increase in density and hardness. The process is expensive, and too tedious to be generally adopted.

A patent has been obtained in this country by M. F. Moll for a method of impregnating wood with creosote by exposing it to the vapour of the *oil of wood-tar*, which is the product of the distillation of the tar.

The first product which passes over when wood-tar is distilled consists for the most part of eupion. When the distillation has been carried so far that the product has about the same specific gravity as water, the receiver is changed, and some lime, or an alkali, is added to the distilled liquid to neutralize the free acid which it contains. On applying a much stronger heat to the tar, impure creosote distils over.

The wood is exposed to the action of the vapours of eupion and creosote in a cast-iron chamber or tank, furnished with some means for applying heat by steam. The wood should be arranged vertically, if convenient; but if not, it should rest on an iron grating, so that the vapour obtains free access to the surface of the wood. Before the timber is exposed to the tarry vapour, the tank is heated to a temperature about 90° or 100° Fahr.; and, after some time, the water expelled from the wood is drawn off, and the vapour of eupion admitted by a pipe from a contiguous boiler. The timber is exposed, in the next place, to the vapour of creosote; and is, lastly, soaked for some time in hot liquid creosote.

The length of time during which the wood should be submitted to these successive operations depends entirely on its hardness and density. As a means of estimating the progress of the different processes, it is recommended to attach to the tank a small test-chamber containing a small piece of the same kind of wood as that in the tank. By observing the progress of the test, that of the large piece may be easily judged by an experienced workman.

This process may be made, without doubt, an effectual means of preserving the wood from decay, but it would seem to be much too complicated and expensive for general adoption on the large scale.

The use of the aqueous solution of creosote for preserving wood has been patented by Mr. Samuel Hall. One hundred parts of water at common temperature dissolve only about 1·25 parts of creosote.

Peroxide and perchloride of tin, and oxide, chloride, and nitrate of copper, are preserving materials, for the use of which a patent was obtained by Mr. Richard Treffry in 1836. To impregnate wood (or any other vegetable material) with oxide of tin or oxide of copper, it is first soaked in a mixture of a pound of quick-lime with about four gallons of water, or else in a solution of a pound of soda-ash (containing about 45 per cent. of alkali) in four gallons of water. When taken out of the alkaline solution, the wood is well washed, and, if convenient, dried. It is next dipped into another tank, containing a solution of either perchloride of tin, chloride of copper, or nitrate of copper. It is immaterial whether the wood is first impregnated with the alkali or the metallic solution, if the superfluous liquid remaining on the surface after the first immersion is carefully removed. The metallic salt preferred by the patentee for wood is chloride of copper, a pound of which may be dissolved in six pounds of water, and a sufficient quantity of the solution used to cover the timber completely. When dry, the timber is ready for use. It is stated that the chloride, nitrate, and acetate of copper may also be applied to wood with advantage by themselves.

The use of a solution of bitumen in oil of turpentine, applied externally as a paint, has been patented by Mr. R. Newton. According to the specification of the patent, the method preferred for making the solution of bitumen is the following:—The bitumen is melted in an iron boiler heated by means of steam, and ten per cent. of common turpentine is added during the melting. When fluid, seventy-five per cent. of oil of schistus, or other mineral oil, is added, the mixture stirred, again heated, and afterwards poured out into an iron vessel to cool. When cold, there are added, first, twenty-five per cent. of common turpentine, and afterwards ten per cent. of hydrate of lime previously sifted and mixed with a small quantity of the liquid. This mixture is said to remain in a permanently liquid state at common temperatures.

The use of a solution of yellow chromate of potash as a preservative agent has been patented by Mr. John Bethell (July, 1838). The bichromate of potash would probably be found a more efficient preservative material than the yellow chromate. Mr. Bethell's patent also includes the application of the refuse lime-water of gas-works; of a solution of caoutchouc in naphtha or turpentine, alone, or mixed with rape-oil, coal-tar, or wood-tar; of a solution of bees'-wax in turpentine, and of drying oil and turpentine. These mixtures are said to impart to the wood both durability and impermeability to water.

The process patented by Sir William Burnett in 1838, for preserving wood and other vegetable matters from decay, consists in impregnating them, in the ordinary manner, with a solution of chloride of zinc, containing one pound of the chloride to five gallons of water. The time required for the digestion of the wood in the solution at common atmospheric pressure varies from ten to twenty-one days, according to the thickness of the wood. Pieces of four inches in thickness, or less, require ten days; pieces of from four to eight inches require fourteen days; and pieces above eight inches require twenty-one days. The timber should be dried in a sheltered situation. It is recommended as an additional precaution that a paint composed of oxide of zinc and drying oil be applied to the wood externally.

The protection from decay afforded to wood by chloride of zinc is said to equal that afforded by corrosive sublimate. Chloride of zinc is better adapted to the preservation of shipping than corrosive sublimate, as the compound which oxide of zinc produces with vegetable albumen is insoluble in sea-water, unlike the compound of oxide of mercury and vegetable albumen. Specimens of English oak, English elm, and Dantzic fir remained perfectly sound for five years in the fungus test-pit at Woolwich (see page 187); but similar unprepared pieces introduced at the same time soon became affected with decay and fungus. The protection afforded to canvass and cordage by chloride of zinc appears to be greater than that by chloride of mercury.

The impure mixture of acetate of peroxide and acetate of protoxide of iron (pyrolignite of iron or dyers' iron liquor), obtained by digesting rusty iron nails, &c. in the crude acetic acid afforded by the distillation of wood, is one of the most economical and efficient of preservative agents. It forms a stable compound with albumen; its acid, when free, exerts no corrosive action on the wood, and being volatile, may be easily expelled from the wood, if necessary, by the application of heat; and lastly, the crude acetate contains a considerable quantity of creosote (page 193). Vegetable matters which easily enter into a state of putrescence, as paste and pulps of carrot and beet-root, are rendered almost inalterable in the air by being soaked in a solution of the crude acetate.

The iron liquor generally employed for preserving wood has the specific gravity 1·056.

To determine the relative amount of protection from decay afforded by the most important of the preceding preservative agents to vegetable matters placed in the same conditions as to moisture and temperature, Dr. Boucherie instituted a set of experiments on wheat flour and pulp of beet-root, of which the following are the results. The experiments, which were all performed at the same time, consisted in mixing equal weights of the vegetable matter, equally moist, with different quantities of the bodies the protective power of which was to be determined.

In all the experiments with wheat flour, 62 grammes were mixed with 30 grammes of water containing the preservative material in solution. A mixture of flour and water only, made for comparison, became completely covered with mould, and evolved a considerable quantity of putrid gas, on the eighth day after being made.

Chloride of mercury.—Three experiments were performed with this substance, in which 2, 4, and 6 decigrammes were dissolved in the 30 grammes of water for mixing with the flour. No alteration had taken place in either of the mixtures at the expiration of two months.

Sulphate of iron.—In five experiments with sulphate of iron, in which from 2 decigrammes to 2 grammes were dissolved in the 30 grammes of water, the appearance of the mould was retarded only a few days. In each mixture it was complete on the twelfth day.

Pyrolignite of iron.—In an experiment in which 1 decigramme of dyers' iron liquor of specific gravity 1·055 (11° Twaddell) was mixed with the usual quantity of flour and water, a slight mould appeared on the tenth day; with 2 decigrammes, on the twelfth day; with 3 decigrammes, on the fifteenth day; with 4 decigrammes, on the twentieth day; with 5 decigrammes and upwards no mould was perceptible up to the sixtieth day.

Arsenious acid.—With 2 decigrammes of arsenious

acid some mould appeared on the thirteenth day; with 4 decigrammes, on the fifteenth day; and with 1 gramme, on the eighteenth day. With 2 grammes no decomposition was perceptible up to the sixtieth day.

Similar results were obtained in experiments with the pulp of beet-root. The decomposition of the pulp was completely prevented by a decigramme of corrosive sublimate; but a gramme and a half of either sulphate of iron, sulphate of copper, or sulphate of zinc only retarded the decomposition of the same quantity of pulp for a few days. A gramme of iron liquor and 6 decigrammes of crude pyroligneous acid were found to be requisite for complete preservation.

§ IV. MODES OF APPLYING PRESERVATIVE AGENTS.

Until lately, the only method commonly practised of conveying a preservative material to the interior of a piece of wood consisted in steeping the wood in a solution of the substance, or else in exposing the wood to the vapour of the preserving body. A billet of wood placed on its end and covered with an aqueous solution gradually absorbs a considerable quantity of the liquid merely by the force of capillary attraction, aided by the pressure of the liquid column. But the impregnation is effected very unequally in this manner, certain parts of the wood presenting far greater facilities for the transmission of the liquid than others. Those parts near the axis, where the tissue is denser than towards the surface, are scarcely at all penetrated by the solution.

The impregnation also takes place with extreme slowness; a piece of wood of about three feet three inches in length, and nine inches in diameter, continued to absorb water and increase in weight after having been submerged in water for ten months.

To obtain a more perfect and rapid impregnation of the wood, Dr. Boucherie* suggested the application of the aspirative force of the tree, the liquid being applied either to the base of the trunk or larger branches, or to the roots. It is indifferent whether the tree is still standing or recently felled. By this force, the liquid is absorbed, in the course of a few days, to a height of eighty or a hundred feet, and even penetrates to the leaves.†

To impregnate a tree recently felled, the base of its trunk may be placed in a vat containing the solution of the preserving material, or else a bag of leather or sheet caoutchouc may be fastened water-tight around the base and put in communication by means of a pipe with a tank or cistern containing

* *Annales de Chimie et de Physique*, t. lxxiv., 113.

† A patent was obtained by Mr. John Bethell for a process for impregnating wood identical in most respects with that of Dr. Boucherie, the butt-end of the recently felled tree being placed in a tank containing the solution, or else the solution is contained in a bag of waterproof cloth affixed to the end of the tree. The process was patented in July, 1838, which was previous to the publication of Dr. Boucherie's paper.

This method of impregnating wood has been favourably reported on by a commission of the French Academy, consisting of MM. Dumas, Boussingault, De Mirbel, Arago, Poncelet, Audouin, and Gambey; and extensive arrangements have been undertaken in France, by the Minister of Marine, for the application of the process to the preservation of wood for the French navy.

the solution. A poplar of about ninety feet in height, the base of which was placed in the month of September in a vat containing a solution of pyrolignite of iron of specific gravity 1·056, absorbed 3 hectolitres (very nearly 10·6 cubic feet) of the solution in the course of six days.

The time which may be allowed to elapse between the felling of the tree and the impregnation varies according to the nature of the tree and the season of the year. At the end of September, a pine, the trunk of which was fifteen inches and a half in diameter, became perfectly impregnated, when put in contact with the solution, forty-eight hours after being felled. In the month of June a plantain was also well penetrated after having been cut down for thirty-six hours. But the sooner the tree is put in contact with the liquid after being cut, the more energetic is the absorption. At the tenth day the aspirative force is hardly sensible.

As the tree should be maintained in a vertical position, its great weight may often become inconvenient to sustain; it is hence sometimes found more advantageous to operate on the tree before it is wholly detached from its roots.

To impregnate a standing tree, two deep notches may be made with a saw on each side of the trunk, into which two narrow wedges are to be inserted to support the tree; or an auger-hole of two or three inches in diameter may be bored through the centre of the tree, and a horizontal cut made by a saw, right and left of the hole, enough of the outside being left to sustain the tree. A bag of tarred leather or sheet caoutchouc is then fastened around the trunk above and below the notches, and placed in communication, by a pipe, with a cistern containing the preserving solution; or else the solution may be contained in a basin of well-tempered clay, large enough to hold two or three gallons of liquid, made around the base of the trunk. To avoid waste of the liquid, the tree may be stripped of its superfluous branches before being submitted to the process. A terminal tuft, however, should always be allowed to remain.

The best season of the year for impregnating the tree, according to the experience of Dr. Boucherie, is the autumn. The impregnation is more difficult to effect in deciduous trees in spring than in winter or summer, but evergreens may be impregnated advantageously in winter.

Different kinds of liquids are not absorbed with equal facility; neutral solutions, for example, are absorbed more readily than either acid or alkaline. A plantain, the trunk of which was about twelve inches in diameter, absorbed in seven days two and a half hectolitres (very nearly 8·8 cubic feet) of a solution of chloride of calcium of specific gravity 1·1095 (about 22° Twaddell).

An objection to the process of impregnating trees by vital absorption is, that it can only be executed in the sap season, which is limited to a few months in the year, and the cutting of the wood at this period is contrary to established practice.

A simpler and equally effective method, by which trees may be impregnated at all seasons of the year, has since been discovered by Dr. Boucherie, and also, independently, by Mr. W. H. Hyett, of Stroud, Gloucestershire, whose Prize Essay on the best solutions for impregnating trees to impart durability, incombustibility, &c., in the Transactions of the Highland Society,* contains a great deal of highly valuable information. The process consists simply in inverting the newly felled tree, stripped of all superfluous branches, divided into convenient lengths, and if necessary, squared, and applying the preserving liquid to the butt-end of the tree, now the uppermost. The liquid may be contained either in a bag of impermeable cloth, adapted to the upper extremity, or in a cup hollowed out of the end of the tree. In most cases, the liqüid quickly penetrates by the superior extremity, and the sap flows out at bottom almost immediately. The operation is terminated when the liquid which issues from the bottom of the piece is the same as that introduced at top. With some woods, which contain a considerable quantity of gas in their pores, the flowing does not commence until the gas is expelled.

It is remarkable that the most porous woods are not those which are most easily penetrated. The poplar resists more than the yoke-elm and the beech; and the willow more than the pear-tree, the maple, and the plane. The ash, according to Mr. Hyett, completely resists the percolation of the liquid.

I am informed by Mr. Hyett that in the month of May every part of the trunks of large beech-trees, with the exception of three or four years'

* Vol. viii. New Series, 1843, p. 535.

growth immediately around the pith, admitted the solution perfectly. At the same season, nine or ten inches in diameter of the heart-wood of Scotch fir-trees of about two feet in diameter resisted the liquids effectually.

The impregnation of timber which has been already seasoned or cut for some time is best attained by first exhausting all its pores of gas, and then introducing the liquid under a considerable pressure. This method was patented by Mr. John Bethell in 1838.

The vessel in which the impregnation is effected is an air-tight iron tank of sufficient strength to withstand an internal pressure of two hundred pounds to the square inch. The circular wrought-iron boilers for high-pressure steam-engines are well adapted for the purpose. The tank is fitted with an air-tight lid or door, and with a common steam-boiler safety-valve, and is connected by one pipe with an exhausting air-pump, and by another pipe with a pressure-pump for forcing the liquid into the pores of the wood. When the wood is introduced into the tank, it is nearly covered with the preserving liquid, and the tank is exhausted of its air. After a short time, air is readmitted, and the liquid forced into the exhausted pores of the wood by the pressure-pump. In some cases, the penetration of the liquid requires to be assisted by applying a gentle heat to the outside of the tank; in others, the liquid enters readily after the exhaustion, without the assistance of pressure. The escape of air from the pores of the wood is expedited by placing the logs of wood in a perpendicular or slanting position, with their top ends above the surface of the liquid.

The apparatus used for injecting wood with a solution of chloride of zinc (Sir William Burnett's patent), at the Portsmouth dock-yard, consists of a cylinder of fifty-two feet in length, and six feet in diameter, capable of containing about nineteen or twenty loads of timber. It is fitted out with a set of exhausting pumps, and a set of pressure-pumps, and has been proved up to 200 pounds to the square inch. When the cylinder is loaded, the air is exhausted to 27·5 inches of mercury, and the liquid is introduced by a pipe in connection with a reservoir. Air is then readmitted and pressure applied, and as the wood absorbs the fluid, the cylinder is again exhausted and the pressure renewed, whereby the fluid is driven into every pore of the wood.*

§ V. OTHER EFFECTS OF THE IMPREGNATION OF WOOD WITH FOREIGN SUBSTANCES.

Besides protection from decay, whether the wood be kept in a dry or humid state, the following effects may be produced by impregnation with certain foreign substances.

1. The increase of the hardness of the wood;

2. The preservation and increase of the flexibility, elasticity, and strength of the wood;

3. The reduction of the combustibility of the wood;

4. The prevention of the expansion and contraction of the wood, and the disjunctions which conse-

* United Service Journal, April, 1843.

quently occur in buildings through variations in the hygrometric condition of the atmosphere;

5. The application of various persistent colours and odours; and

6. The increase of the density of the wood.

1. From the effects of wood prepared with pyrolignite of iron (page 198) on cutting tools, its hardness has been estimated by workmen at double that of the unprepared wood.

Of some specimens of beech impregnated by Mr. Hyett, a carpenter considered that with acetate of copper to be the hardest; those with common salt, yellow prussiate of potash, sulphate of copper, and corrosive sublimate, to be next in hardness; and those with pyrolignite of iron, sulphate of iron, and nitrate of soda, next. Of some specimens of prepared larch the hardest was that with pyrolignite of iron; the next in hardness were those with sulphate of iron and corrosive sublimate; and the next, those with acetate of copper, sulphate of copper, and prussiate of potash.

2. The flexibility and elasticity of wood may be preserved any length of time, according to Dr. Boucherie, by slightly impregnating the wood with some deliquescent substance, as a dilute solution of chloride of calcium or chloride of magnesium, by which a certain degree of humidity is always preserved in the wood, if exposed to the atmosphere. The solution preferred by Dr. Boucherie as the most economical is the mother-liquor of the salt-works, which contains small quantities of each of the above chlorides. The flexibility and elasticity are stated to be in

proportion to the quantity of saline matter introduced. A plate of pine-wood charged with the mother-liquor, of three millimetres (·118 inch) in thickness and sixty centimetres (23·6 inches) in length, was capable of being bent into three concentric circles without being broken, and when allowed, would again become straight. Its flexibility and elasticity were found to be undiminished after the lapse of eighteen months.

Wood which contains a small quantity of chloride of calcium or chloride of magnesium does not become dry by exposure to the sun in the middle of summer, and the little moisture lost by the wood during the day is again absorbed at night. The adherence of paints and resinous varnishes does not seem to be affected by the application of these deliquescent substances.

The mother-liquor of salt-works would of itself tend to preserve the wood from decay; for security, however, it is recommended to add to the solution about a fifth part of the pyrolignite.

But Mr. Hyett has been led to conclude, from his experiments, that the flexibility of wood does not depend in all cases on the presence of moisture. Pieces of larch impregnated with acetate of copper and sulphate of copper were found to be far more flexible than a piece impregnated with chloride of calcium. To ascertain the flexibility and strength of wood impregnated with different substances, three specimens of each tree were planed down to an inch square, till they passed as accurately as possible through a gauge, and cut to the length of four feet. The lengths were then placed horizontally in a frame so constructed that a weight suspended from the middle could not vary its position from the irregular bending of the piece; the ends were supported on props three feet apart. The weights were applied as marked in the following tables every half-minute, and the deflection at the end of the interval, and the breaking point, were noted for each weight. The tables shew the mean of the three observations for each piece.

I. DEFLECTION OF LARCH, IN INCHES.

Weight applied in pounds.	Chloride of calcium. No. 13.*	Sulphate of iron. No. 19.*	Sulphate of copper. No. 20.*	Corrosive sub-limate. No. 21.*	Acetate of copper. No. 22.*	Pyrolignite of iron. No. 23.*	Natural state. No. 24.*	Prussiate of potash. No. 25.*
28	·43	·26	·63	·33	·51	·28	·53	·36
56	·83	·63	1·5	·73	1·23	·65	1·38	·88
84	1·36	1·11	3·3	1·6	2·68	1·11	4·6	2·03
98	1·73	1·48		2·06	5·03	1·5		
105	2·05	1·8				1·81		
112	2·43	2·18				2·65		
118	2·85	2·85						
124	3·73	3·58						
128	4·7							

* The numbers refer to the table, pages 218, 219.

II. DEFLECTION OF BEECH, IN INCHES.

Weight applied; lbs.	Corros. sub. No. 1.*	Nitr. soda. No. 2.*	Pruss. pot. No. 3.*	Pyrol. iron. No. 5.*	Chlor. sodium. No. 6.*	Sulph. iron. No. 7.*	Sulph. cop. No. 8.*	Acet. cop. No. 9.*	Nat. state. No. 17.*	Nat. state. No. 26.*	Nat. state. No. 28.*
56	·33	·51	·06	·26	·4	·26	·36	·33	·38	·25	·4
112	·71	1·8	·33	·63	1·16	·83	1·1	·86	·84	·56	·8
140	·98		·5	·85		1·2	1·8	1·26	1·16	·75	1·05
154	1·15		·6	1·0		1·41	2·9	1·55	1·37	·85	1·2
161	1·26		·65	1·1		1·55		1·76	1·49	·9	1·25
168	1·41		·71	1·18		1·73		2·0	1·66	·95	1·35
174	1·51		·76	1·26		1·88		2·21	1·79	1·01	1·4
180	1·66		·81	1·36		2·06		2·48	1·95	1·05	1·5
184	1·78		·83	1·45				2·76	2·1	1·1	1·55
188	1·86		·88	1·51				3·1	2·27	1·13	1·6
192	1·98		·91	1·6				3·6	2·48	1·16	1·65
196	2·1		·96	1·66				3·98	2·69	1·2	1·7
200	2·26		1·00	1·76						1·23	1·75
204	2·41		1·05	1·85						1·28	1·85
208			1·08	1·95						1·33	1·9
212			1·11	2·06						1·36	2·0
216			1·15	2·16						1·43	2·0
220			1·2	2·3						1·46	2·1
224			1·26	2·41						1·48	2·15
229			1·31	2·6						1·53	2·25
233			1·4	2·78						1·61	2·35
238			1·5	3·01						1·68	2·5
243			1.6							1·78	2·6
247			1·68							1·85	2·75
252			1·8							1·93	3·0
257			1·96							2·03	3·3
261			2·05							2·13	
266			2·21							2·23	
270			2·4								
274			2·58								
279			2·83								
283			3·16								
288			3·48								

* The numbers refer to the table, pages 218, 219.

From the results of Mr. Hyett's experiments, contained in the preceding tables, it appears that the strength of the wood may be greatly increased or diminished by impregnation with foreign substances, and that it is most diminished by those substances which tend most to preserve or increase the flexibility of the wood. In the case of beech, the greatest deflection with a weight of 112 pounds is produced by nitrate of soda, chloride of sodium, and sulphate of copper; but the pieces impregnated with nitrate of soda and chloride of sodium were the first to break, being unable to support a weight of 140 pounds; the piece with sulphate of copper broke next, under a weight of 161 pounds. On the other hand, the piece of beech which shewed least deflection with a given weight, namely that impregnated with prussiate of potash, was the strongest, and able to support the weight of 288 pounds.

It is to be observed that the flexibility and strength of larch and beech are not affected in a similar manner by the same substance, but the experiments on both kinds of wood lead to the conclusion that those prepared pieces which are deflected most by a given weight are those which are broken soonest on increasing the weight, and the reverse.

The preceding tables also lead to the important conclusion that the two different classes of trees, resinous and non-resinous, require very different treatment. In the beech, and probably all other non-resinous trees, prussiate of potash and pyrolignite of iron are the only agents which do not impair the strength of the wood in its natural state; while in the larch, prussiate of potash and sulphate of copper are the only substances which do not increase the strength of the wood. By far the greatest strength is imparted to beech by prussiate of potash; on larch, the same agent produces no alteration. Sulphate of iron diminishes the strength of beech, but considerably increases that of larch. Sulphate of copper and acetate of copper also diminish the strength of beech, but not that of larch.

For beech, the sulphates of iron and copper are not so beneficial as the corresponding acetates; this circumstance may be referred to the corrosive action which sulphuric acid exerts on woody fibre, especially on that of trees which do not contain any resin. Acetic acid exerts no such corrosive influence.

Corrosive sublimate produces much the same effect on larch as on beech. The pyrolignite of iron may be considered the best single material to be applied to both kinds of trees, but prussiate of potash is decidedly the best for beech, and chloride of calcium the best for larch.

3. The reduction of the inflammability and combustibility of the wood is not the least important of the effects attainable by impregnation with saline substances, especially common salt, chloride of calcium, and chloride of magnesium. Not only is the inflammability of the wood diminished, but its combustion, when fairly commenced, is rendered difficult by the access of air to the carbonized wood being impeded by the thin film of fused alkaline or earthy salt.

Two huts, one built of prepared wood, and the other of unprepared, were set on fire at the same time by applying equal weights of the same lighted combustible matter. When the hut built of ordinary wood had become reduced to ashes, the interior surface of the other had hardly become carbonized (Dr. Boucherie). If perfectly dry, there appears to be little or no difference between the inflammability of prepared and unprepared wood.

4. The expansions and contractions which wood often experiences through changes in the hygrometric state of the atmosphere, and the consequent loosening of joints which thereby occurs, may also be prevented or diminished by impregnation with some deliquescent substance. According to Dr. Boucherie, wood containing a small amount of moisture is not subject to these changes in volume, and they may be entirely prevented by a little chloride of calcium or chloride of magnesium. A few large thin tables made of wood thus prepared underwent no change in form or size during a twelvemonth, while similar tables in the same situation, made of unprepared wood, became exceedingly warped. The addition of a little pyrolignite of iron to the deliquescent substance is also recommended, to ensure durability.

5. The colours which are most easily applied to wood by the aspirative process are those which are produced by double decomposition between two substances in solution, the respective solutions being introduced into the wood consecutively. Thus, to produce a blue tint, the wood may be first impregnated with a solution of yellow prussiate of potash, and afterwards with a solution of persulphate of iron;

or the same solutions may be applied in the reverse order. The tint in this case is derived from Prussian blue. A black tint may be imparted by introducing successively a solution of sulphuret of sodium and a solution of acetate of lead, whereby sulphuret of lead is produced. Wood may also be stained black by introducing an infusion of galls and pyrolignite of iron. A green (Scheele's green) may be applied by means of acetate of copper and arsenious acid; a reddish brown (prussiate of copper), by sulphate of copper and yellow prussiate of potash; and a delicate yellow (chrome yellow), by acetate of lead and bichromate of potash. A solution of sulphate of copper, to which a slight excess of ammonia has been added, penetrates the wood with facility, and produces an agreeable blueish tint.

As the impregnation is not effected equally through the whole substance of the wood, the tinting is not uniform, but in veins and waves, which present an agreeable appearance when the wood is worked up and polished.

According to Mr. Hyett, different solutions do not penetrate the same parts with equal facility. In applying acetate of copper and prussiate of potash to larch, it was observed that the sap-wood was coloured most, and the heart-wood least, when the acetate was introduced first. But when the prussiate was first applied, the heart-wood became most deeply coloured. With sulphate or acetate of copper first, and prussiate of potash next, beech may be made to appear very much like mahogany. Iodide of lead and iodide of mercury cannot be applied to wood with advantage as colouring materials.

Pyrolignite of iron alone produces in beech a dark grey colour, from the action of the tannin contained in the wood on the oxide of iron; but in larch and Scotch fir it merely darkens the natural colour of the wood. Prussiate of potash alone produces a dingy green colour. The tints of most of these colouring materials, especially of the prussiates of iron and copper, are improved by exposure to light; and the richest colours are obtained when the process is rapidly executed. (Mr. Hyett.)

Vegetable colouring matters do not easily penetrate the wood by the aspirative process, probably on account of the affinity of the woody fibre for the colouring principle, whereby the whole of the latter is abstracted from the solution by those parts of the wood with which it is brought at first into contact.

Essential oils and other odoriferous matters may be easily introduced into the wood in a state of solution in weak alcohol; and the odours thus imparted are considered to be as durable as those supplied by nature. Wood may also be impregnated with resinous substances in alcoholic solution, by which it may be rendered impervious to water, and far more inflammable.

6. The increase which wood experiences in density by being impregnated with foreign substances is shewn in the following tables drawn up from the experiments of Mr. Hyett.

SPECIFIC GRAVITY OF PREPARED BEECH.

No. in Table, pages 218 *and* 219.	*Preparation.*	*Specific gravity.*
No. 28.	Dried 2 years over stove	681
„ 26.	Soaked in water 2 years and dried in air 3 years	763
„ 2.	Nitrate of soda	809
„ 7.	Sulphate of iron	832
„ 17.	Natural state (green)	840
„ 1.	Corrosive sublimate	844
„ 5.	Pyrolignite of iron	859
„ 8.	Sulphate of copper	866
„ 3.	Yellow prussiate of potash	876
„ 6.	Chloride of sodium	888
„ 9.	Acetate of copper and vinegar	937

SPECIFIC GRAVITY OF PREPARED LARCH.

No. in Table, pages 218 *and* 219.	*Preparation.*	*Specific gravity.*
No. 24.	Natural state (green)	488
„ 20.	Sulphate of copper	533
„ 21.	Corrosive sublimate	541
„ 25.	Yellow prussiate of potash	551
„ 22.	Acetate of copper and vinegar	552
„ 13.	Chloride of calcium	560
„ 19.	Sulphate of iron	595
„ 23.	Pyrolignite of iron	608

The most important results of Mr. Hyett's experiments on the best materials for impregnating wood are embodied in the following table.

RESULTS OF EXPERIMENTS ON

By Mr. W. H. Hyett.

Experiment.			Tree.				Material applied.				
No.	Date of commencement.	Duration in days.	Sort.	Age in years.	Height in feet.	Diameter in inches.		No. of pounds in gallon of water.	Quantity absorbed in gallons.	Cost. s.	d.
1	May 3	24	Beech	56	55	6½	Corros. sub.	⅛	16½	4	1½
2	,,	22	,,	56	45·5	6½	Nitr. soda	1	12	3	3
3	,,	23	,,	50	48¼	5½	Pruss. pot.	½	8	9	0
4	,,	30	,,	imperfect			Imp. gas-tar	...	...	...	
5	May 5	19	,,	56	34	6½	Pyrolig. iron	...	7½	6	3
6	,,	21	,,	56	42	6½	Com. salt	1	14	0	4
7	May 6	19	,,	56	54½	6½	Sulph. iron	½	17	2	10
8	July 13	17	,,	56	28½	7	Sulph. cop.	1	17¼	8	7½
9	,,	17	,,	56	34½	8½	Acetate cop. & vinegar	¼ 1 pint	17½	14	2
10	May 13	20	Larch	12	32	5	Gas-tar	...	1	0	3½
11	,,	20	Scotch fir	12	29	5	Do.	...	1	0	3½
12	,,	12	,,	12	22	4	Corros. sub.	1/18	9½	2	4½
13	May 15	17	Larch	12	34	5	Chlor. of cal.	¼	9	3	0
14	May 14	15	Lime	12	21	3½	Corros. sub.	1/18	9½	2	4½
15	,,	14	Elm	12	18	3	Do.	1/18	7½	1	10½
16	,,	14	White poplar	12	24½	4½	Do.	1/18	10	2	6
17	June 2	...	Beech	56	37	9	Nat. state	...	...	...	
18	,,	...	,,	56	...	...	Sulph. iron	Saturated solution	...	...	
19	July 13	17	Larch	12	34	6	Do.	½	12½	2	1
20	,,	17	,,	12	35½	7	Sulph. cop.	1	15	7	6
21	,,	17	,,	12	35	5	Corros. sub.	18	10¾	2	8
22	,,	17	,,	12	29	6	Acetate cop. & vinegar	¼ 1 pint	13¾	11	0
23	,,	17	,,	12	29½	6	Pyrolig. iron	...	1	0	10
24	July 30	...	,,	12	36	7	Nat. state	...	...	...	
25	July 13	17	,,	12	33	7	Pruss. pot.	½	13¾	13	9
26	,,	...	Beech	large old timber, cut 5 years, soaked in water containing much carbonate of lime for 2 years, and afterwards dried in an open shed.							
27	,,	...	Pear	cut 5 years and perfectly dried over a stove.							
28	,,	...	Beech	cut 2 years and perfectly dried over a stove.							

MATERIALS FOR PRESERVING WOOD.
(*Trans. of the Highland Society.*)

Observations on the state of preservation of the prepared woods after having remained for nine months in a damp cellar surrounded with decaying saw-dust full of fungi.
Very clean and dry, except where the solution had not touched, where there are fungi.
Covered with fungi, not very damp.
Tendency to fungi and decay.
Fungi, particularly on parts where the tar abounds.
Clean and dry.
Very wet, black, a little fungi.
Dry and clear, except where the iron did not reach.
Tendency to decay.
Dry, much fungi.
Tendency to fungi and decay.
Dry and clean, with little or no fungi.
Light, dry, and clean.
Tendency to decay.
A little fungi.
Much fungi.
Dry and clean.
Decay and fungi.
Fungi and mouldiness.
Clean where solution had touched, but not elsewhere.
Clean, but a little fungi.
Dry and clean.
Dry, with a little fungi.
Clean and dry, except on the parts untouched by the solution.
Clean and dry.
Light and dry, but a little fungi.
Wet, tendency to decay and fungi.

DYEING AND CALICO-PRINTING.*

§ I. History of Dyeing and Calico-printing.—II. General Properties of Vegetable Colouring Matters.—III. General Nature of Dyeing Processes.—IV. Calico-printing Processes.

In the various operations of dyeing and calico-printing are exhibited some of the most refined and ingenious applications of chemical science. Though many processes in these arts were practised for ages before any just views were entertained of the chemical nature of tinctorial substances, yet dyeing is strictly a chemical art, and it cannot be properly understood without some acquaintance with the chemical properties of the acting bodies.

The great object of all dyeing operations is the impregnation of a textile fabric with coloured substances derived from animals, vegetables, and minerals, in such a manner as to render them incapable of being removed by washing with water. The modes of effecting this object vary as greatly as the colouring matters differ from each other in their chemical habitudes. Though the chemical re-actions which are exhibited in the various dyeing and print-

* For a considerable portion of the materials from which the present article is compiled, I am indebted to Mr. Mercer, of the Oakenshaw print-works, near Blackburn, and to Mr. John Graham, of the Mayfield print-works, Manchester.

ing processes are, for the most part, sufficiently intelligible, yet they are sometimes of a highly complex character; and the theoretical principles of a few valuable processes, discovered accidentally, are even yet but imperfectly understood.

§ I. HISTORY OF DYEING AND CALICO-PRINTING.

In the East Indies, in Persia, in Egypt, and in Syria, the art of dyeing has been successfully practised from time immemorial. In the books of the Pentateuch frequent mention is made of linen cloths dyed blue, purple, and scarlet, and of rams' skins dyed red; and the works of the tabernacle, and the vestments of the high-priest, were enjoined to be of purple.

The place of antiquity where dyeing was most extensively carried on, as the general business of the inhabitants, was probably Tyre, the opulence of which city seems to have proceeded in a great measure from the sale of its rich and durable purple. This colour was prized so highly, that in the time of Augustus a pound of wool dyed with that material cost, at Rome, a sum nearly equal to thirty pounds of our money. The Tyrian purple is now generally believed to have been derived from two different kinds of shell-fish, described by Pliny under the names *purpura* and *buccinum*, and was extracted from a particular organ in their throats to the amount of one drop from each fish. It is at first a colourless liquid, but by exposure to air and light becomes successively citron-yellow, green, azure, red, and, in the course of forty-eight hours, a brilliant purple. If the liquid is evaporated to dryness soon after being collected, the residue does not become coloured in this manner. The purple is remarkable for its durability; it resists the action even of caustic alkalies and most acids. Plutarch observes in his Life of Alexander, that, at the taking of Susa, the Greeks found in the royal treasury of Darius a quantity of purple stuffs of the value of five thousand talents, which still retained its beauty, though it had lain there for one hundred and ninety years. The properties of the colouring juices of shell-fish have been investigated by Cole, Gage, Plumier, Duhamel, and Reaumur, who have succeeded in procuring a purple dye, though inferior to what may be obtained by other dye-stuffs.

It does not appear that the art of dyeing was much cultivated in ancient Greece. In Rome it received a little more attention; but very little is now known of the processes followed by the Roman dyers, such arts being held, by them, in too little estimation to be considered worth describing. The principal ingredients used by the Romans were the following: — of vegetable matters, —alkanet, archil, broom, madder, nutgalls, woad, and the seeds of pomegranate and of an Egyptian acacia: of mineral productions,—copperas, blue vitriol, and a native alum mixed with copperas.

The progress of dyeing, as of all other arts, was completely arrested in Europe, for a considerable time, by the invasion of the northern barbarians in the fifth century. In the East the art still continned to flourish, but it did not revive in Europe until towards the end of the twelfth, or the beginning of the thirteenth century. One of the principal places where dyeing was then practised was Florence, where it is said there were no less than two hundred dyeing establishments at work in the early part of the fourteenth century. One of the Florentine dyers having ascertained, in the Levant, a method of extracting a colouring matter from the lichens which furnish archil, introduced this material into Florence on his return; by its sale he acquired an immense fortune, and became one of the principal men of the city.

The discovery of America tended greatly to the advancement of the art, as the dyers became supplied from thence with several valuable colouring materials previously unknown in the old world; amongst which are logwood, quercitron, Brazil-wood, cochineal, and annatto. A great improvement in dyeing also took place about the year 1560, which consisted in the introduction of a salt of tin as an occasional substitute for alum. With cochineal, the salt of tin was found to afford a colour far surpassing in brilliancy any of the ancient dyes. The merit of this application is attributed to Cornelius Drebbel, a Dutch chemist, whose son-in-law established an extensive dye-house at Bow, near London, about the year 1563.

About the middle of the sixteenth century, logwood and indigo began to be employed in Europe as dyes, but not without considerable opposition from the cultivators of the native woad. The use of logwood was prohibited, in England, by Queen Elizabeth, by a very heavy penalty, and all found in the country was ordered to be destroyed. Its use was not permitted in England till the reign of Charles the Second. Indigo, one of the most valuable and important of dye-stuffs, was also forbidden to be used in England and on the continent, and denounced as " food for the devil."

These, and similar prejudices, were gradually surmounted, and in the eighteenth century the art of dyeing made very considerable progress. Madder, from which the colour known as Turkey or Adrianople red is produced, then began to be properly appreciated; and quercitron, a fine yellow dye, was brought extensively into notice by Dr. Bancroft. But the chief improvements of the moderns in this art consist in the employment of pure mordants, and in the application of colours derived from mineral compounds, as peroxide of iron, Prussian-blue, chrome-yellow, chrome-orange, manganese-brown, &c. Each of these colouring matters may be obtained as an insoluble precipitate on mixing together two solutions: in the dyeing processes the proper solutions are made to mix and produce the precipitate within the fibre, by impregnating it first with one solution, and afterwards with the other. As the precipitate thus produced is imprisoned within the fibre, it is not removable by mere washing with water.

The mode of dyeing Turkey red, which is the most durable vegetable colour known, was discovered in India. It was afterwards practised in other parts of Asia and in Greece; and, about the middle of last century, dye-works for this colour were established near Rouen and in Languedoc by some Greek dyers. In 1765, the French government, convinced of the importance of the process, caused an account of it to be published; but it was not introduced into this country until the end of last century, when a Turkey-red dye-house was established in Manchester by M. Borelle, a Frenchman. M. Borelle obtained a grant from Government for the disclosure of his process, but the method which was published does not seem to have been very successful. A better process was introduced into Glasgow about the same time by another Frenchman, named Papillon. Previous to this, however, Mr. Wilson of Ainsworth, near Manchester, had obtained the secret from the Greeks of Smyrna, and published it in two essays, read before the Literary and Philosophical Society of Manchester; but the process was said to be expensive, tedious, and less applicable to manufactured goods than to cotton in the skein. The greater part of the Turkey-red dyeing executed in this country is still carried on in the Glasgow district.

The ancients seem to have attained considerable proficiency in the art of topical dyeing, or of producing coloured patterns on cloths. The variegated linen cloths of Sidon are noticed by Homer, who lived nine hundred years before Christ, as very magnificent productions. In India the art of imparting a coloured pattern to a cotton fabric has been practised with great success from time immemorial, and it derives its English name of calico-printing from Calicut, a town in the province of Malabar, where it was formerly practised on a very considerable scale. According to Herodotus, who wrote more than four hundred years before Christ, the inhabitants of Caucasus adorned their garments with representations of various animals by means of an aqueous infusion of the leaves of a tree; and the colours thus obtained were said to be so fast as to be incapable of being removed by washing, and as durable as the cloth itself. The material of which the cloths were made is not stated, but they were probably woollen, as that part of Asia was then, as at present, celebrated for the superior quality of its wool.

From the following account by Pliny of the nature of the process of topical dyeing practised by the ancient Egyptians, it would appear that this people had attained such proficiency in the art, as could only have been originally acquired by extensive practice and close observation.

"An extraordinary method of staining cloths is practised in Egypt. They there take white cloths and apply to them, not colours, but certain drugs which have the power of absorbing or drinking in colour; and in the cloths so operated on there is not the smallest appearance of any dye or tincture. These cloths are then put in a cauldron of some colouring matter, scalding hot, and after having remained a time are withdrawn, all stained and painted in various colours. This is indeed a wonderful process, seeing that there is, in the said cauldron, only one kind of colouring material; yet from it the cloth acquires this and that colour, and the boiling liquor itself also changes, according to the quality and nature of the dye-absorbing drugs which were at first laid on the white cloth. And these stains or colours, moreover, are so firmly fixed as to be incapable of being removed by washing. If the scalding liquor were composed of various tinctures and colours, it would doubtless have confounded them all in one on the cloth; but here one liquor gives a variety of colours, according to the drugs previously applied. The colours of the cloths thus prepared are always more firm and durable than if the cloths were not dipped into the boiling cauldron." (Pliny, Hist. Nat., lib. xxxv. cap. 11.) In as few words the principle of the common operations of calico-printing could hardly be more accurately described.

The *pallampoors*, or large cotton chintz counterpanes made in the East Indies from a very early period, have similar dye-absorbing drugs applied to them by the pencil, and certain parts of the cloth are coated with wax to prevent the absorption of colour when immersed into the vessel containing the dye.

The topical dyeing of cotton goods seems to have been practised for a considerable time in Mexico. When Cortez conquered that country, he sent to Charles V. cotton garments with black, red, yellow, green, and blue figures. The North American Indians have also been for a long time in possession of a mode of applying patterns in different colours to cloth.

The art of calico-printing does not appear to have been much practised in Europe until the close of the seventeenth or the beginning of the eighteenth century, when Augsburg became famous for its printed cottons and linens. From that city the manufacturers of Alsace and Switzerland were long supplied with colour-mixers, dyers, &c. The first print-ground in England was founded by a Frenchman on the banks of the Thames near Richmond, and soon afterwards a more considerable one was established at Bromley Hall in Essex. Several others were some time afterwards founded in Surrey, in order to supply the London shops with chintzes, the importation of which from India had been prohibited by an act of parliament passed in 1700, on account of the excessive clamours of the silk and woollen weavers. Though merely intended as a protection to the English silk and woollen manufacturers, this act had the effect of greatly stimulating and increasing the infant art of calico-printing; for the demand for printed calicos and chintzes could then be gratified only by printing, in this country, white Indian calicos, the importation of which was still allowed under a duty. An excise duty of threepence per square yard was imposed on the printed calicos in 1712, which was increased to sixpence in 1714; but the importation of calico being still considerable, a new alarm was raised, and a law enacted in 1720, which prohibited the wearing of all printed calicos whatever, whether of foreign or home production. The operations of the printer were then confined to the printing of linens.

The oppressive and absurd act of 1720 was repealed in 1730; but the calicos then permitted to be printed were to have the warp of linen and merely the heft of cotton, and were subject to a duty of sixpence per square yard. With such discouragements, the progress made in calico-printing was extremely slow: so lately as the middle of last century it was computed that only fifty thousand pieces of the mixed cloth were printed annually in the whole of Great Britain; whereas, at the present time, several manufacturers turn out as much as three and four hundred thousand pieces per annum each. The part of the act of 1730 by which the warp was required to be made of linen yarn was repealed in 1774; but the printed calicos were still subject to a duty of threepence-halfpenny per square yard, the repeal of which in 1831 has been of the utmost advantage both to the manufacturer and the consumer.

The wonderful developement which calico-printing has received within the last half-century is to be attributed, in a great measure, to the adaptation of numerous ingenious mechanical inventions. The improvement in patterns, and the reduction in the price of cotton prints during this period, are striking illustrations of the advancement which has been made in machinery. The first improvement on the original wooden hand-printing block,* which is quite similar to the block of a wood-engraving, consisted in the substitution, for some styles of work, of copper plates, about three feet square (similar to those employed for printing engravings on paper), on which a much more delicate pattern could be engraved than on wood. The colour being laid on the copper plate and the superfluous colour removed by a thin steel scraper, the plate was passed with the

* A description of the various modes of printing cloths now practised will be found in another part of the present article.

cloth through a press similar in principle to that of a copper-plate printer. The engraving of the plate was executed either by a common graver or by a punch.

The greatest mechanical improvement ever effected in this art was the invention of cylinder or roller printing, which is said to have been first made by a calico-printer at Jouy in France, named Oberkampf, in whose hands alone it remained for some time. The invention appears also to have been made independently by a Scotchman of the name of Bell, and was first successfully applied in the large way about the year 1785, at Monsey near Preston. Cylinder printing has received its greatest developement in Lancashire; and the perfection to which the process has been there brought is the chief cause of the admitted superiority of our calico-printing establishments over those on the Continent, where cylinder printing is comparatively but little practised.

Printing by the cylinder is executed with not only greater accuracy than by the wooden block, but with a saving of time and labour almost incredible. One cylinder machine, attended by one man to regulate the rollers, is capable of printing as many pieces as a hundred men and a hundred girls could with the hand-block during the same time; or as much work may be executed by a cylinder machine in four minutes as by the ordinary method of block-printing in six hours. A length of calico equal to one mile has been printed off with four different colours in a single hour.

The successful application of an engraved copper cylinder was followed by that of a wooden roller having the pattern in relief, the mode of printing by which is known as "surface printing." The "union" or "mule machine," which is a combination in one machine of the engraved copper cylinder with the wooden roller in relief, was invented about 1805 by Mr. James Burton of Church, near Blackburn.

One of the most important of recent improvements in the mechanical department of calico-printing is a very ingenious method of executing block-printing with several colours by press-work in an arrangement similar to one of the modern type-printing machines. (An account of this mode of printing is contained in another section of the present paper.) Another important modern improvement, more particularly adapted to the press-machine, consists in the substitution of stereotype blocks made of a mixed metal, tin, lead, and bismuth, in the place of the wooden block.

During the last century the chemical principles of dyeing and calico-printing were investigated by Dufay, Hellot, Bergmann, Macquer, and Berthollet, and numerous and valuable improvements were suggested by some of their researches. The application of chlorine, by Berthollet, to the bleaching of tissues, especially cotton and flax, contributed in no small degree to the advancement of these arts; it is during the present century, however, and from the researches of numerous chemists still living, that they have received the most essential assistance from chemistry. The chief improvements introduced by the moderns consist, as already observed, in the application of colours derived from mineral substances.

Among the earliest introduced of this class of bodies, were iron buff and Scheele's green, which were followed by antimony orange (first applied by Mr. Mercer) and Prussian blue. The two chromates of lead (chrome-yellow and chrome-orange) were next introduced by M. Kœchlin of Mulhausen in 1821, and a few years afterwards Mr. Mercer first applied, on the large scale, the peroxide of manganese, known as manganese bronze.

§ II. GENERAL PROPERTIES OF VEGETABLE COLOURING MATTERS.

By far the greater number of the colouring matters employed in the art of dyeing are derived from vegetables, but the animal and mineral kingdoms also contribute a small number. Colouring principles are abundantly distributed over all the organs of vegetables, but never in a state of purity; they are always mixed, more or less, with other substances, and their isolation in a pure state often requires very complicated processes. Only a small number, comparatively, of these substances have as yet been obtained sufficiently pure to have their chemical composition determined. Like almost all other vegetable principles, they are composed either of carbon, hydrogen, and oxygen, or else of the preceding elements together with nitrogen, and have received particular designations derived in general from the names of the plants by which they are furnished. The most common colour of the vegetable kingdom is green, but as the substance which gives rise to this colour in leaves and trees is of an unctuous nature, it cannot be easily applied to cloth: to obtain a green, the dyer generally has recourse to the admixture of a yellow with a blue colouring matter. It is remarkable that the most vivid and brilliant of vegetable colours, namely, those of flowers and other parts of the plant exposed to solar light, are so small in quantity, and so fugitive, that they are of all the most difficult to isolate. In the organs which are protected from the light, as the interior of stems, branches, and roots, the colouring matters are generally devoid of all brilliancy, but when separated from the accompanying substances, they exhibit considerable lustre, and are by far the most durable.

Nearly all the colouring matters of plants which are capable of being isolated are yellow, brown, and red; the only blue substances which have been procured from plants are indigo and litmus, and no black vegetable substance, strictly speaking, has ever been isolated.

As a particular class of bodies, vegetable colouring matters do not possess many chemical characters in common; they are associated rather on account of their common application in the arts, than from the possession of similar properties. Most of them are entitled to be ranked among acids, but others are strictly neutral. By far the greater number of them are soluble in water, and always in larger proportion in hot than cold water. Those which are insoluble in water generally dissolve in alcohol, ether, and fixed oils.

In dry air, vegetable colouring matters appear to be permanent, but in humid air, and especially under the influence of the solar rays, they gradually lose colour, and become converted, by the absorption of oxygen from the air, into yellowish brown or colourless compounds. The ultimate action of air or oxygen on organic colouring matters in the presence of moisture, is to convert their carbon into carbonic acid, and their hydrogen into water. Solutions of organic colouring matters in water are acted on by oxygen with far greater facility than the dry colours.

The colour of some of these bodies is changed in a very remarkable manner by the application of acids and alkalies. The blue colour of most flowers, that of the flowers of the violet, for instance, is rendered red by acids, and green by alkalies. The purple infusion obtained by boiling red cabbage in water is affected in a similar manner; acids produce with it a lively red, and alkalies a full green. If the dried petals of the red rose are digested in spirits of wine, or hot water, they lose their colour without affording any, or at most only a trace of colour to the liquid. On adding a few drops of sulphuric acid, however, the liquid immediately acquires a fine red colour, and, if a slight excess of an alkali is afterwards added, it becomes green. The change from red to green may be produced indefinitely. The purple colour of litmus is rendered red by an acid, and blue, not green, by an alkali.

Some animal and vegetable colouring matters must undoubtedly be regarded as neutral bodies, that is, as possessing neither the characters of an acid nor those of a base; but most of them, particularly such as are soluble in water, have all the essential characters of a weak acid, being capable of uniting with and neutralizing salifiable bases, as potash, soda, lime, magnesia, alumina, &c. This tendency to combination is not confined, as some have supposed, to soluble colouring matters and insoluble bases; but the union is more obvious in such cases, as the resulting compound is always insoluble, while soluble bases usually form soluble compounds with soluble colouring matters.

For alumina and certain metallic peroxides, especially peroxide of iron and peroxide of tin, some organic colouring matters possess an energetic attraction. The pigments commonly called *lakes* are insoluble compounds of colouring matters with alumina or oxide of tin, which may be formed by mixing a solution of alum or of perchloride of tin* with the infusion of the dye-stuff, and adding afterwards an alkaline carbonate to liberate peroxide of tin or alumina: as the latter precipitates, it unites with and carries down the colouring matter in solution, frequently leaving the supernatant liquid entirely colourless. The infusion of the dye-stuff is sometimes made with an alkaline liquor and mixed with a solution of alum after being filtered. In this way *yellow lake* is made with a decoction of turmeric, and annatto and quercitron lakes from the respective dye-stuffs.

Important applications are made in dyeing and calico-printing of the attraction which exists between alumina and metallic peroxides on the one hand, and

* Commonly called spirits of tin.

organic colouring principles on the other. By first impregnating a piece of cloth with alumina, green oxide of chromium, peroxide of iron, or peroxide of tin, and then dipping it into the infusion of the dye-stuff, the colouring matter leaves the solution to unite with the base, forming an insoluble compound, whereby it becomes strongly attached to the tissue, and is rendered less susceptible of alteration by the air, the solar rays, and other decomposing agents.

The attractive force of colouring matters for insoluble bases has been regarded by some as a mere attraction of surface, analogous to, if not identical with, the force of cohesion or adhesion, being the same as the attractive power by which charcoal is enabled to withdraw colouring substances from their solutions, and also the same as that by which a solid body condenses a permanent gas upon its surface. This mechanical attraction, which always exerts itself between a solid on the one hand, and a substance in solution or a gas on the other, depends entirely on the state of the surface of the solid, and is in no way connected with the chemical relations of the combining substances.

But the combinations of alumina, &c. with soluble colouring matters seem to be cases of true chemical combination, taking place in definite proportions, and under the influence of different degrees of attractive force for different colouring principles. Thus, alumina has a stronger attraction for the colouring principle of madder than for that of logwood, and a stronger attraction for that of logwood than for that of quercitron. When a piece of cloth impregnated with alumina is immersed in a decoction of quercitron bark, it acquires a fast yellow colour; if the same cloth is washed for some time and kept in a hot decoction of logwood, the alumina parts with the colouring principle of quercitron to combine with that of logwood, and the colour of the cloth becomes changed from yellow to purple. If the same cloth is next immersed for a few hours in a hot infusion of madder, the alumina parts with the colouring principle of logwood to unite with that of madder, the colour of the cloth changing from purple to red. The quantity of alumina on the cloth does not appear to diminish while these substitutions are taking place. These interesting facts were communicated to me by Mr. John Thom of the Mayfield print-works.

By contact with chlorine, and in presence of a little moisture, the colour of most, but not all, vegetable and animal dye-stuffs is instantly destroyed; the organic substance is decomposed, being commouly converted into colourless products, from which the original colour cannot be reproduced by any known process. In at least one case, however, which is that of indigo, the colour is reproducible after having been discharged by chlorine, provided the quantity of chlorine applied to the indigo has been no more than sufficient to change the blue colour to a buff, and not enough to destroy all colour. The rich crimson colour into which some preparations of indigo are changed by chlorine is also convertible into blue, though not to so deep a shade as the original indigo (Mr. Mercer).

In most cases of the destruction of vegetable colours by chlorine, the decomposition is effected, without doubt, through the powerful affinity of chlorine for hydrogen, which may be manifested in two ways; 1st, in the direct abstraction of hydrogen from the organic substance, and 2ndly, in the decomposition of water, the hydrogen of which unites with the chlorine to form hydrochloric acid, while the oxygen of the water decomposes the colouring matter, forming carbonic acid with its carbon, and water with its hydrogen. Chlorine does not bleach readily in the absence of all moisture, and hydrochloric and carbonic acids may generally be discovered among the products. In a few cases, however, the bleaching action of chlorine simply consists in the direct combination of the chlorine with the colouring matter to form a compound which is devoid of colour.

Chromic acid is another powerful bleaching agent, which acts by affording oxygen to the colouring matter, becoming itself reduced to the state of green oxide of chromium. The colour of the vegetable substance is even more readily destroyed than if chlorine had been applied.

Most vegetable colouring matters are also bleached by sulphurous acid in the presence of water. The action of this substance is not so energetic as that of chlorine, and differs from it essentially in the circumstance that the colours are not entirely destroyed, but may in general be restored by exposure to the air, or by the application of a stronger acid or an alkali.

It is uncertain whether the bleaching power of sulphurous acid depends on the partial deoxidation of the colouring matter, or on the union of the sulphurous acid with the colouring matter to form a colourless combination.

Charcoal has also been classed among bleaching agents, as it readily withdraws colouring matters from their solutions, frequently leaving the supernatant liquid entirely colourless. The charcoal which absorbs colouring matters with most avidity is that obtained by the calcination of bones and other animal matters, the superiority of which seems to depend merely on its minute state of division, whereby the contact of the liquid and charcoal is rendered more perfect. The action of charcoal in bleaching vegetable infusions is altogether different from that of chlorine, and also from that of sulphurous acid. The colouring matter is not decomposed, but is merely mechanically attached to the surface of the charcoal, without having experienced any chemical alteration whatever.

When brought into contact with deoxidizing agents, several organic colouring matters part with a portion of their oxygen, and at the same time lose their colour. But if afterwards exposed to the air, or any source of free oxygen, the deoxidized bodies reassume oxygen, and with that element their original colour. The coloured bodies would therefore appear to be compounds of oxygen with a colourless radical. The alternate reduction and oxidation may be practised on the same substance indefinitely. As examples of colouring matters susceptible of these changes, may be mentioned litmus, logwood, Brazil-wood, sapanwood, peachwood, red beet-root, and the red cabbage. The most convenient deoxidizing agents to be employed in such experiments are the following:

1°. A mixture of granulated or feathered tin and a caustic alkali.

2°. Protoxide of iron, or protoxide of tin, recently precipitated, and still moist.

3°. Hydrogen gas, applied in the nascent state, by introducing a piece of zinc or iron into the infusion of the colouring matter, rendered acid by the addition of muriatic or sulphuric acid.

4°. Sulphuretted hydrogen gas, a stream of which may be passed through the coloured infusion, or the latter may be agitated in a jar containing the gas. As the colour disappears, a whitish precipitate of sulphur is produced.

5°. Double metallic sulphur salts containing an alkaline sulphuret, such as the sulphuret of arsenic and potassium (sulpharsenite of potash).

It is worthy of note that the colourless or white radicals of Brazil-wood, logwood, sapanwood, &c. do not unite with alumina and metallic peroxides to form insoluble compounds or lakes, like their oxides, or the true colouring matters.

Other vegetable colouring principles than those above mentioned become converted into colourless substances when exposed to the action of deoxidizing agents, but the chemical change which some of them suffer appears to be the acquisition of hydrogen instead of the yielding up of oxygen. When exposed to the air or other source of free oxygen, this hydrogen is removed, and the original colour returns. Indigo is one of the colouring matters susceptible of such changes.

Several colouring principles are contained in the plants from which they are derived in their white, deoxidized, or hydruretted state. Such is the case, for example, with indigo. That substance does not exist in the blue or dehydruretted state in the plant by which it is furnished, but as white indigo, or indigotin, the colourless hydruret of indigo-blue. Most vegetable juices, the recent pulp of fruits, detached leaves, &c. become coloured brown and yellow by exposure to the air, from the absorption of oxygen. If carefully kept in a vessel of some gas devoid of free oxygen, such bodies experience no change in colour.

Colouring matters have the property of uniting with animal and vegetable tissues, by virtue of an attraction of surface quite similar to that by which they unite to animal charcoal. When well-scoured wool or silk is digested in a decoction of cochineal, logwood, or Brazil-wood, or a solution of sulphate of indigo, it abstracts the colour so completely as to leave the liquid colourless, as if animal charcoal had been introduced. The affinity of vegetable tissues for colouring matters is in general not so great as that of animal tissues for the same substances. The vegetable fibre readily combines with a colouring material; but unless the latter is insoluble in water, the combination is exceedingly feeble. A familiar example of the affinity of the vegetable fibre for organic colouring matters is presented in the staining of a linen napkin by red wine. The portion of the cloth on which the wine falls soon abstracts the whole colour from the liquid, becoming dyed red; but beyond the spot thus produced, the cloth becomes moist without acquiring an appreciable colour, the wine having been deprived of all its colour by the portion of the cloth with which it came first into contact. The attractive force by which this result is obtained must not be considered as peculiarly subsisting between tissues and organic colouring matters, as many mineral substances are withdrawn from their solutions by tissues in quite a similar manner. Thus, cotton cloth readily separates lime from lime-water,* and the insoluble sulphate of alumina from an aqueous solution of basic alum.

Vegetable and animal colouring principles are divisible into two classes, with reference to their solubility or insolubility in water. Those which are soluble readily attach themselves to tissues, but only with a feeble affinity, as they may be separated by continued washing in water, especially with the assistance of heat. Logwood, madder, Brazil-wood, cochineal, and in fact, the greater number of dye-stuffs, belong to this class. To unite them firmly to a tissue, another substance is applied, which possesses the property of forming an insoluble combination with the colouring matter. Those colouring matters which are of themselves insoluble, or but slightly soluble in water, generally form, as might be expected, much faster combinations with tissues. Indigo, annotta, safflower, and such yellow and brown dyes as contain tannin combined with substances of the nature of apothème, are the principal members

* The separation of lime from lime-water by cotton cloth is exhibited when a drop of a solution of bleaching powder (which always contains free lime) is allowed to fall on a piece of cotton dyed with indigo. On the spot where the solution first touches the cloth, the colour remains unaltered, the lime only having been there intimately absorbed ; but on the ring surrounding this spot, the colour becomes discharged through the action of the chloride of lime.

of this class. To effect their solution, some other solvent than pure water must be applied. Thus, indigo is dissolved by bringing it, through the action of a deoxidizing agent, to the state of white indigo or indigotin, which is soluble in water in the presence of an alkali or some lime. If a piece of cloth is dipped into such a solution, the white indigo is absorbed into the pores of the fibres, and on exposing the cloth to the air, imbibes oxygen, by which it becomes converted into the original insoluble indigo blue. The latter remains firmly attached to the fibre, being imprisoned within the pores, and therefore incapable of being removed by mere washing in water. The colouring matters of annotta and safflower, though very sparingly soluble in water, are easily dissolved by alkaline liquids, from which they may be precipitated on the addition of an acid. A piece of silk might be dyed with either of these colours, by first impregnating it with the alkaline infusion of the dye-stuff, and then passing it through a weak acid: the best method, however, of dyeing both silk and cotton with annatto or safflower is by wincing the piece in an imperfectly neutralized alkaline infusion of the dye-stuff, which contains the colouring matter in a state of feeble suspension, readily precipitated on a solid body presenting a finely divided surface, such as cloth. The partial neutralization of the alkali in this process is effected by a very weak acid or an acidulous salt, such as bitartrate of potash (cream of tartar).

Such are the principal general properties of organic colouring principles, a knowledge of which is of the highest importance to the practical dyer and calico-printer. But these bodies differ so much from each other in many respects, that the best means of extracting them from the organs by which they are produced, and the most effectual manner of applying them to textile fabrics, can only be discovered by the accurate investigation of the chemical and general properties of each distinct colouring matter separately.

Nature of colour.—The appearance of colour may almost be regarded as an optical delusion, since it does not exist in the object, but in the light which the object reflects. It is well known that a ray of white light from the sun is resolvable into three rays of the primary colours, red, yellow, and blue. As a mixture or combination of these three coloured rays, white or colourless light may be considered, the absence of colour depending on the exact equilibrium of the three. When the coloured rays are partially separated by the refractive force of a glass prism, an image or spectrum is obtained, presenting seven different colours, namely, red, orange, yellow, green, blue, indigo, and violet. The orange, green, indigo, and violet tints proceed from the intermixture in various proportions of the three primary rays.

Among opaque substances, there are some which completely absorb the three coloured rays incident on their surface, and therefore, having no light to reflect to the eye, appear *black*. Others, on the contrary, reflect all the light, and are consequently *white*. But others possess the power of decomposing the light, that is, of absorbing the whole or a portion of one of the three primary rays and refleeting the remainder; or, it may be, of absorbing unequal proportions of each of the three rays. When such is the case, the body appears to be coloured, not from the inherent possession of a colour, but because the light which it reflects to the eye is not homogeneous white light. A blue substance, for example, is said to reflect the blue rays only, or in greatest proportion, the yellow and red rays being absorbed. If red, it is said to absorb the yellow and blue rays, and reflect the red; and by the absorption of the rays in unequal proportions, and by the reflection of more or less of the white or undecomposed light, every shade of colour may be produced. The same remarks apply to transparent coloured substances; only, instead of the decomposed light being reflected to the eye, it is transmitted. According to this manner of viewing the colouring principle, it has been observed that the art of dyeing consists in fixing upon stuffs, by means of molecular attraction, substances which act upon light in a manner different from the stuffs themselves.

The production of white by the combination of the three primary colours is practised in one of the finishing operations to which goods are subjected in the process of bleaching. To whatever length the ordinary operations may be continued, some kinds of goods always retain a brownish-yellow hue, which may be removed, and a pure white imparted, by applying a little smalts, indigo, archil, or a mixture of Prussian-blue and cochineal pink. In such cases the blue, or mixture of blue and pink, supplies the tints necessary to the production of white with the brownish-yellow colour of the goods. But when the dyer attempts to form white by combining red, yellow, and blue, he often obtains a dark brown, or black, because the resulting combination does not reflect as much light as the three coloured ingredients separately.

The following alphabetical list of colouring matters, with their origin, uses, and principal chemical characters, may prove useful for reference. The history and applications of some of them will be fully discussed in separate articles.

I. LIST OF VEGETABLE AND ANIMAL COLOURING MATTERS.

Alkanet.—The root of the *Anchusa tinctoria.* Its colouring principle, which is red, is nearly insoluble in water, but soluble in alcohol, ether, oil of turpentine, and fixed oils. It is used as a colouring matter for ointments and other unctuous preparations, but not in dyeing.

A variety of alkanet was formerly met with in commerce, derived from the roots of the *Lawsonia inermis.*

Annotta.—A hard paste prepared by inspissating the washings from the fermented seeds of the *Bixa orellana.* Its colouring matter is yellowish-red, nearly insoluble in water, soluble in alcohol and alkaline liquids. It forms an orange-coloured compound with alumina, a citron-yellow compound with protoxide of tin, and a greenish-yellow compound with protoxide of copper. It is used to dye silks golden-yellow, by simply digesting the goods in an alkaline solution of annotta, and orange-red by exposing them afterwards to the action of a dilute acid. It is also used to dye cotton yellow, with the aluminate of potash as the mordant, and as a colouring matter for cheese.

Archil.—A violet-coloured paste, made from different species of lichens: that of the Canaries, which is the most esteemed, is from the *lichen rocellus;* and that of Auvergne, from the *lichen parellus. Litmus*, *turnsole*, and *cudbear* are merely modifications of archil. The colouring principle of these dye-stuffs is soluble in water and alcohol, and its colour is changed by the weakest acids from purple or violet to bright red. It is a brilliant colour, but possesses little permanence, and is chiefly used to give a violet or purple bloom as a finish to silks and woollen cloths already dyed with other colours. It is rarely used for cotton goods.

Barwood.—This is a dull red dye-stuff, the colouring matter of which is only slightly soluble in water, but sufficiently so for dyeing without the application of another solvent, such as an alkaline liquid, in which it dissolves with facility. It gives red compounds with alumina and peroxide of tin, and is mostly used for dyeing silks and woollen cloth. The colouring matter of *camwood* is quite similar in its properties to that of barwood, but is somewhat brighter in colour. Both barwood and camwood possess much more permanence than peachwood, for which they are now frequently used as substitutes.

Brazil-wood.— This and *Sapanwood*, *Fernambouc-wood*, *Peachwood*, and *Nicaragua-wood*, are derived from certain species of Cesalpina. Their colouring matter, which seems to be identical, is red, soluble in water, rendered purple or blue by alkalies, and yellow by acids. It forms a red compound with alumina, a black compound with peroxide of iron, a violet compound with protoxide of tin, and a rose-coloured compound with peroxide of tin. It is of itself a fugitive colour, being easily bleached by light with exposure to the air; but its stability is considerably increased by being combined with peroxide of tin or alumina. It is used in dyeing wool, silk, and cotton with the tin and aluminous mordants. Of these woods peachwood and sapanwood are the most extensively employed at present.

Camwood.—(See Barwood.)

Catechu, or *terra Japonica.*—This is an extract from the heart-wood of the khair-tree of Bombay and Bengal (*mimosa catechu*), made by evaporating the decoction of the wood nearly to dryness. Its chief constituent is a variety of tannin, differing slightly in its characters from that contained in galls. Catechu is very soluble in water and alcohol, with the exception of a little earthy matter. It gives a rich brown-grey colour with nitrate of iron, a fast bronze by being oxidized through the agency of a mixture of sulphate or nitrate of copper and muriate of ammonia, a brownish-yellow with protochloride of tin, and a reddish-brown with acetate of alumina. It is extensively used in calico-printing as a topical brown, when mixed with nitrate, sulphate, or acetate of copper, and sal-ammoniac.

Cochineal.—A female insect found on the *cactus opuntia* or *nopal*, dried. Its colouring principle, termed *coccinellin*, is naturally of a purplish-red colour; it is soluble in water and weak alcohol; its colour is changed to red by acids, and to crimson by alkalies. It forms a fine crimson compound with alumina, a violet compound with protoxide of tin, and a scarlet compound with peroxide of tin. Wool and silk are dyed of a fine scarlet by means of a mixture of decoction of cochineal with cream of tartar and dyers' spirit, which is a mixture of protochloride and perchloride of tin; and of a crimson, by a decoction of cochineal with alum and perchloride of tin. Cochineal is also used in the preparation of the pigment called carmine.

Cudbear.—(See Archil.)

French berries, called also *Avignon berries* and *Persian berries.* The fruit of the *rhamnus infectorius.* The berries afford a bright yellow liquid when boiled in water, which gives a golden-yellow colour with protochloride of tin, a lemon-yellow with peroxide of tin, a rich yellow with alumina, and a drab with a salt of iron. They are much used as a bright yellow topical colour when combined with a tin or aluminous mordant.

Fustet or *yellow fustic.*—The wood of the *rhus cotinus*, the colouring matter of which is yellow. Being a fugitive dye-stuff, it is very little employed at present in dyeing processes.

Fustic or *old fustic.*—The wood of the *morus tinctoria.* Its aqueous decoction is orange-coloured, and is brightened by cream of tartar, alum, and solution of tin. Its principal use is to dye woollen and cotton cloths of a permanent yellow with an aluminous mordant. It is also used to produce a brownish tint with copperas.

Indigo.—A blue insoluble pigment procured by the oxidation of a colourless substance (indigotin) contained in the leaves of the *indigofera*, by subjecting the leaves to a process of fermentation. Indigo blue may be reconverted to indigotin by applying a deoxidizing agent, and then becomes soluble in alkaline liquids, in which form it may be applied to cloth (see page 244).

Kermes grains. — Dried female insects of the species *coccus ilicis*, which are found on the leaves of the *quercus ilex* or prickly oak. The decoction of kermes in water is red, and is rendered brownish by acids, and violet by alkalies. Kermes was formerly much used as a crimson dye with a mordant of alum, but it is now superseded by cochineal and lac-dye.

Lac-dye. — *Stick-lac* is an exudation produced by the puncture of an insect on the branches of several plants, by which the twig becomes incrusted with a brownish-red resin. This is a complicated mixture, containing a small portion of a red colouring matter quite similar to that of cochineal. *Lac-dye* is the residue of the evaporation of the aqueous infusion of ground stick-lac. It is employed to dye wool of a brilliant scarlet colour with a mordant of dyers' spirit. The solution of the lac for this purpose is effected in very dilute muriatic acid.

Litmus.—(See Archil.)

Logwood.—(Campeachy wood.)—The wood of the *hæmatoxylon campechianum.* Though the colouring principle of logwood is red in its natural state, yet it forms blue or violet compounds with almost all metallic oxides. It is soluble in water, affording a reddish liquid, which is rendered purple by alkalies, or, if added in excess, brownish-yellow. It is employed in dyeing all kinds of stuffs of a variety of shades between light purple and black with an aluminous mordant, and between lilac and black with the acetate of iron as the mordant.

Madder.—Dutch madder is the root of the *rubia tinctorum*, and Turkey and French madder that of the *rubia peregrina.* According to M. Runge, madder contains five distinct colouring principles; madder red (called also *alizarine*), madder purple, madder orange, madder yellow, and madder brown. Madder red is soluble in water, but only in small proportion, and therefore cannot be employed in a concentrated solution. It is very extensively used in the dyeing and printing of cotton goods for the production of a permanent bright red colour with an aluminous mordant; of a lilac, purple, and black with oxide of iron; and of a variety of shades of chocolate with a mixture of the iron and aluminous mordants, with or without the addition of sumach. Turkey madder is preferred for producing the Turkey-red dye, pinks, and light lilacs; and Dutch madder for producing purples, chocolate, and black. A form of madder containing more colouring matter than the natural root is now met with in commerce, under the name of *garancine.* This article is said to be prepared by digesting powdered madder in cold oil of vitriol, which destroys most of the constituents of the root, but leaves the red colouring matter unaltered.

Nicaragua-wood.—(See Brazil-wood.)

Peachwood.—(See Brazil-wood.)

Quercitron.—The bark of the *quercus nigra*, or

yellow oak, which grows in North America. Its colouring principle, which is yellow, is very soluble in water. Quercitron is much used to impart a yellow colour to cotton with the intervention of an aluminous mordant, and to produce drabs with an iron mordant, and olives with a mixture of the iron and aluminous mordants. It is also much used, when mixed with a small quantity of madder, to produce an orange with a mordant of alumina.

Safflower.—The flowers of the *carthamus tinctorius.* Safflower contains two colouring matters; a yellow substance soluble in water, which is of no value in dyeing; and a fine red substance, insoluble in water, but soluble in an alkaline liquid. It is used to dye silk and cotton of a rose colour by wincing the piece in an imperfectly neutralized alkaline infusion of the dye-stuff.

Sandal-wood. (Red Sanders wood.)—The wood of the *pterocarpus santalinus.* Its colouring matter is red, scarcely soluble in water, but soluble in alkaline lyes. The colour may be applied to tissues by dipping them alternately in an alkaline decoction of sandal-wood and in some acidulous liquid. It does not stand exposure to light well.

Sapanwood.—(See Brazil-wood.)

Sumach, *galls*, *valonia*, and *sawwort.*—These and some other astringent vegetable productions are used to impart to cloth a variety of shades from slate colour to black, with peroxide of iron as the mordant. These matters can hardly be classed among colouring matters, as their active ingredients, tannic and gallic acids, are white when pure. By uniting with peroxide of iron, however, these acids form blueish-black compounds, which are the basis of common writing-ink, and may be communicated to cloth by first boiling the piece in a decoction of the astringent material, and afterwards digesting it in a solution of copperas. An infusion of logwood is commonly added to the solution of copperas. Vegetable astringent principles are also used in some other dyeing processes, in which their action seems to partake of that of a mordant.

Turmeric.—The root of the *curcuma longa.* Its colouring principle, which is orange-yellow, is slightly soluble in water, and readily soluble in an alkaline solution, becoming dark brown. As a dye, it is applied only to silk.

Turnsole.—(See Archil.)

Weld.—The entire dried plant, *reseda luteola.* Its decoction in water is yellow. Silk, woollen, and cotton goods may be dyed of a permanent and bright yellow by a decoction of weld with alumina or peroxide of tin as the mordant. With the latter, weld affords to cloth the fastest vegetable yellow colour we possess.

Woad.—The colouring matter of this plant (*isatis tinctoria*) seems to be identical with indigo. Woad is commonly employed as a fermentative addition to indigo in the pastel vat.

II. LIST OF MINERAL COLOURS EMPLOYED IN DYEING.

Antimony orange.—This orange-red substance has been applied to cloth by passing the piece through a solution of the sulphuret of antimony and a little sulphur in a caustic alkali, and afterwards exposing it to the air to precipitate the sulphuret, through the absorption of carbonic acid.

Arseniate of chromium.—This is a fine grass-green coloured compound, which may be imparted to cloth, by the application, first of a solution of chloride of chromium, and afterwards of a solution of arseniate of soda.

Chrome-yellow, or *chromate of lead.*—The colour of this pigment is bright yellow; it may be communicated to cloth by the consecutive application of solutions of acetate or nitrate of lead and bichromate of potash; or the oxide of lead may be first fixed on the cloth in an insoluble state, as carbonate, tartrate, or sulphate. It consists of one equivalent of chromic acid and one equivalent of oxide of lead.

Chrome-orange, or *subchromate of lead.*—This is a dark orange-red pigment, consisting of one equivalent of chromic acid and two equivalents of oxide of lead. To apply it to cotton, the piece is first dyed with chrome-yellow, and is afterwards passed through hot milk of lime, by which a portion of the chromic acid of the chrome-yellow is separated.

Manganese brown (*hydrated peroxide of manganese*).—Cloth is dyed with this substance by being passed, first, through a solution of sulphate or chloride of manganese; next, through a caustic alkaline solution, to precipitate protoxide of manganese; and lastly, through a solution of chloride of lime, to convert the protoxide of manganese into peroxide; or the peroxidation may be effected by mere exposure to air.

Orpiment (*sulpharsenious acid*).—This is a bright but alterable yellow, which may be communicated to silk, wool, and cotton, by first passing the goods through a solution of orpiment in ammonia, and afterwards suspending them in a warm atmosphere to volatilize the ammonia and precipitate the orpiment. This substance is sometimes applied in the form of a solution in a caustic fixed alkali, in which case the precipitation is afterwards effected by passing the cloth through dilute sulphuric acid.

Peroxide of iron (iron buff). — This oxide is applied to cloth to produce a yellowish-brown shade of different intensities, by passing the piece through a solution of a salt of the peroxide of iron, and a solution of an alkaline carbonate, in succession.

Prussiate of copper.—A delicate cinnamon colour is sometimes communicated to cotton by means of this substance, which is applied by first passing the cloth through a solution of sulphate of copper, then through a dilute alkali to precipitate oxide of copper, and lastly, wincing in a solution of yellow prussiate of potash, containing a little muriatic acid.

Prussian blue.—To apply this pigment, the cloth may be first impregnated with a solution of acetate of iron (iron liquor), and afterwards passed through a solution of yellow prussiate of potash, acidified with a little muriatic acid.

Scheele's green (arsenite of copper).—This grass-green coloured substance may be applied to cloth by the double decomposition of nitrate of copper and arsenite of potash; the cloth being passed through solutions of these salts consecutively. A better method is, first to precipitate oxide of copper on the cloth by the action of an alkali, and to wince the piece afterwards in a solution of arsenite of potash.

§ III. GENERAL NATURE OF DYEING PROCESSES.

The processes by which different kinds of textile fabrics are impregnated with the same colouring material are often very dissimilar, and few dyeing processes are applicable in their details to goods of cotton, silk, and wool. For this reason, the observations in the present section refer chiefly to cotton fabrics, the treatment of which requires greater assistance from chemistry than the more easily dyed animal tissues. The dyeing of cottons, however, is mostly practised as a part of the process of calico-printing; but the chemical principles involved in the different operations are precisely the same, whether the cloth is merely dyed and finished in that state, or both printed and dyed.*

Fig 30.

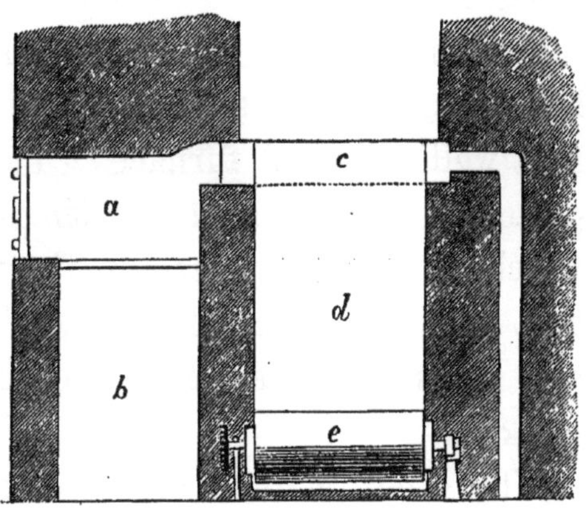

The object of the first operation to which cotton goods are subjected, whether intended to be afterwards printed or merely dyed, is the removal of the fibrous down or nap on the surface of the cloth. This is effected by the process of *singeing*, which may be performed in two different ways equally efficaci-

* For detailed accounts of distinct dyeing processes, the reader is referred to other articles treating of individual colouring materials.

S

ous. The old method consists in drawing the cloth swiftly over a red-hot semi-cylindrical bar of copper, three-quarters of an inch in thickness, placed horizontally over the flue of a fire-place, situated immediately at one end of the bar. The disposition of the different parts of a singeing furnace may be understood with the assistance of the sectional representation in fig. 30; *a* represents the fire-place, and *b* the ash-pit; *c* is the semi-cylindrical bar of copper, forming the top of the flue; and *d* is the strip of cotton, which is rapidly drawn over the ignited bar, and immediately passed round a wet roller, *e*, to cool from the effects of singeing. In the figure, the flue is represented as passing downwards to communicate with the common draught-chimney.

An iron bar, of two inches or more in thickness at the top, was formerly used instead of a copper bar; but the latter is found to last about ten times as long as the former, and to singe nearly three times as many pieces of cotton with the same consumption of fuel. With a copper bar, about fifteen hundred pieces may be singed by a ton of coals in a well arranged furnace. The cotton is generally passed over the bar three times; twice on the side which is to be printed, or the "face," and once on that which is to be the "back." By this operation the colour of the calico becomes very similar to nankeen.

The other method of singeing consists in passing the cloth rapidly through a coal-gas flame, for which a patent was obtained by Mr. Hall of Basford, near Nottingham, in the year 1818. The gas issues from numerous perforations through the upper surface of a horizontal tube, and the cloth to be singed is drawn over the flame rapidly by rollers. In the method first patented, the flame is drawn up through the web of cotton or other fabric by a flue leading into a common draught-chimney; but the draught not being always sufficient to draw the flame through immediately, an improvement in the apparatus was devised by Mr. Hall, and patented in 1823, which consisted in placing immediately over the gas-flame a horizontal tube, with a slit lengthways through its lower surface, which tube is placed in communication with a fan or an exhausting apparatus. An arrangement of this kind, so constructed as to allow the passage of two pieces of cloth at the same time over two gas-flames, is capable of singeing, when properly managed, fifty pieces per hour.

That the colours of the tinctorial matters applied to tissues may appear in their purity, it is essential that the cloth be wholly freed from the foreign matters which adhere to its surface, whether imparted in the processes of spinning, weaving, &c., or else naturally adherent to the fibre of the cloth. In cotton goods, this is accomplished by the process of bleaching by means of chlorine; and in silk and woollen goods, by the action of sulphurous acid.

The ordinary operations practised in the process of bleaching by chlorine consist in subjecting the cloth to the successive action, 1°, of a dilute alkaline solution; 2°, of a dilute solution of chloride of lime or bleaching powder (commonly called "chemic" in bleach-works and print-works); and 3°, of dilute sulphuric acid. The operation of submitting the cloth to a solution of bleaching powder is known as "chemicking;" and to dilute sulphuric acid, as "souring." The action of sulphuric acid on the cloth impregnated with a solution of bleaching powder is to liberate chlorine by combining with the lime to form sulphate. The details of this important process will form the subject of another article; for the present, the following list of the successive operations to which a cotton fabric is subjected in order to prepare it for printing, will suffice for purposes of reference:

1. Washing in cold water,
2. Soaking for eight hours in boiling lime-water,
3. Washing in cold water,
4. Souring,
5. Washing,
6. Soaking for ten hours in a dilute solution of soda-ash,
7. Washing,
8. Chemicking,
9. Souring,
10. Washing,
11. Soaking in solution of soda-ash,
12. Washing,
13. Chemicking,
14. Souring,
15. Washing,
16. Soaking in hot water,
17. Squeezing and drying.

In the process of bleaching mousselin de laines by means of sulphurous acid, the goods are usually passed two or three times through a solution of soap, and soda, at about the temperature 130° Fahr., and then exposed for several hours to the action of sulphurous acid gas, produced from burning sulphur in a close chamber. The latter operation is termed "sulphuring." The goods are next passed through a very weak solution of caustic soda, dried, and usually impregnated with a dilute solution of tin, which imparts considerable brilliancy to the colours afterwards applied to the goods. For this purpose, de laines (which are formed of cotton and wool) are impregnated with two different solutions of tin consecutively, one intended to afford oxide of tin to the wool, the other to the cotton. The solution first applied is a mixture of perchloride of tin and muriatic acid, for the wool; the other is stannate of potash,* from which oxide of tin is precipitated on the cotton by passing the piece afterwards through dilute sulphuric acid. For the finer work, the sulphuring of de laines is usually performed twice.

The only operations to which silken cloth is subjected preparatory to being printed, are, 1°, boiling in a solution of soap and soda to remove the "gum;" 2°, passing through dilute sulphuric acid; and 3°, washing and drying.

To impart a permanent dye to a tissue, it is essential that the colouring material, or the substances from which it is to be produced, should be applied in a state of *solution*, or in a condition to penetrate to the interior of the fibre of the cloth, either at its open extremity, or through the parietes. If a

* A solution of oxide of tin in caustic potash.

piece of cloth is dipped into common writing-ink, the black colour acquired by the cloth may be removed by washing with water, because the tannate of iron, which is the basis of the ink, instead of being in solution, is in an insoluble state, being merely suspended in the liquid, and therefore unable to enter the interior of the fibre. To apply the tannate of iron in a permanent manner, it is necessary to *produce it within the fibre*, which is accomplished by first imbuing the cloth with an infusion of galls or other liquid containing tannin, and afterwards with a solution of a salt of iron. That the colours in solution in the dye-beck should attach themselves to the stuff in the form of a compound insoluble in their original solvent, is the principle on which the dyeing of fast colours rests; and the more insoluble the compound in other liquids, so much the faster the colour.

In nearly all the different processes for dyeing cloths, the colour is applied by one of the four following methods:

1. From two solutions; the colouring material not existing in either separately, but produced on the mixture of the two. The cloth is first impregnated with one solution, and afterwards with the other.

2. From the solution of the colouring material; the cloth being first impregnated with some substance (usually existing on the cloth in the solid state), which has the property of combining with the colouring matter to form an insoluble compound.

3. From the solution of the colouring material

itself, or its basis; the cloth having previously undergone no essential preparation.

4. By effecting a chemical alteration of the fibre of the cloth, with the formation of a coloured product.

By the first method of dyeing, which is the simplest and most intelligible, all mineral colours, such as chrome-yellow, Prussian-blue, peroxide of iron (iron buff), and manganese-brown, may be applied to textile fabrics. The proper colouring matter in all these cases is insoluble in water, and is thrown down as a precipitate whenever the two solutions proper for its formation are mixed. Thus, whenever an aqueous solution of the salt called bichromate of potash is mixed with an aqueous solution of acetate of lead, an insoluble precipitate of chrome-yellow (chromate of lead) is produced. In like manner, Prussian-blue is precipitated when a solution of yellow prussiate of potash is mixed with a solution of a salt of the peroxide of iron. In the processes of dyeing cloth with these and all other mineral colours, the mixture of the proper solutions, and consequent formation of precipitate, is made to take place within the elongated cell or tube which forms the fibre of the cloth; so that the resulting solid, being imprisoned within the fibre, is rendered incapable of being removed by mechanical means. The fastness of colours applied to cloth in this way is entirely a mechanical effect, and in no way referable to a chemical attraction of the fibre for the colouring matter.

A piece of white cotton cloth moistened with either

a solution of bichromate of potash or of acetate of lead, may be easily cleaned from every particle of the soluble bichromate or acetate by simply washing with water. But if the piece of cloth is first imbued with the solution of the acetate, and afterwards with the bichromate, (or the order of impregnation may be reversed,) the precipitate of chrome-yellow is produced within the fibre, and can never be removed by washing with water. The chrome-yellow that is washed away in this experiment was merely loosely attached to the exterior of the fibre. It may be observed here, that as all the colouring matter which is deposited on the exterior of the fibre is a loss of material, it is advantageous to remove the excess of the solution with which the cloth is first imbued, by draining, squeezing, and sometimes by slightly washing the cloth when taken out of the first solution and about to be immersed in the second.

So far as the mere operations are concerned, the processes now commonly practised for applying mineral colours to cloth, are rather referable to the second style of dyeing according to the classification here adopted. Instead of passing the cloth through the two solutions consecutively, one of the two materials, usually the base of the mineral colouring matter, is first applied in an insoluble state, and the cloth is afterwards winced or agitated in a *dilute* solution of the other. To apply ferro-prussiate of copper, for example, to cotton in this way, the piece is first impregnated with a solution of sulphate of copper. The oxide of copper is, in the next place, fixed in an insoluble state by passing the cloth through a dilute alkaline solution; and the prussiate of copper is formed, lastly, by wincing the cloth in a dilute solution of yellow prussiate of potash containing a little muriatic acid. The economization of the solution last applied is the chief advantage of such a mode of proceeding over the old method of applying the two solutions in succession, without the fixation of a substance derived from the first; since the production of any superfluous colouring material is entirely avoided. In the process of dyeing cotton with chrome-yellow, the same depth of colour may be imparted to a piece containing precipitated oxide or carbonate of lead, from a solution of eight ounces of bichromate of potash to a gallon of water, as from a solution of twenty-four ounces of bichromate in a gallon of water, when the cloth contains only the soluble acetate or nitrate. It is to be observed, that although the mere operations in this kind of work are the same as those of the second style to be noticed immediately, yet the principles of the two styles are dissimilar, for the colouring material is not contained in the second solution any more than in the first.

To apply to cloth in a permanent manner those colouring substances which are naturally soluble in water, and have not in themselves a strong affinity for tissues (see page 243), of which kind are the greater number of vegetable and animal tinctorial matters, it is essentially necessary to effect their conversion on the cloth into compounds which are insoluble in water. This is accomplished by first applying to the cloth some substance (most frequently the sub-salt of a metallic oxide) which has an affinity for the colouring matter, whereby it is enabled to withdraw it from the solution and convert it into an insoluble compound. The substance which thus acts as the bond of union between the tissue and the colouring matter, is distinguished as the *mordant*. One circumstance in which this style of dyeing differs from the preceding is, that in this, the mordant must be applied to the cloth before the colouring matter, except in some cases where both may be applied at the same time; but with mineral colours, which may be imparted by the successive application of two solutions, it is generally a matter of indifference which of the two solutions the cloth is first impregnated with.

In its common acceptation by the practical dyer the term mordant is as indefinite as it is inappropriate, since it includes any kind of substance which can facilitate the application of a dye-stuff to a tissue. Properly speaking, a mordant is a substance which has an attraction of surface for the tissue, a chemical affinity for the colouring matter in solution (see page 236), and the property of forming an insoluble compound with the colouring matter. By virtue of the combination of these properties, it is enabled to effect in a durable manner the union of the tissue with the colouring substance.

But with the practical dyer, the term mordant has a much wider signification; even the solvent of the dye-stuff, if the latter is insoluble in water, receives that designation; thus sulphuric acid is sometimes termed a mordant when employed as a solvent for indigo in the preparation of Saxon blue. The name, which was given by some French dyers at a time when little was known respecting the chemical principles of dyeing, is derived from *mordere*, to bite; the mordant being supposed to exert a corrosive action on the fibre which serves to expand the pores and allow the absorption of the colour.

In most cases of cotton dyeing with the intervention of a mordant, the latter must exist on the cloth about to be dyed in a form insoluble in water. But as it is also essential that the mordant should be contained in the interior of the fibre of the cloth, it must be applied at first in a state of solution, for no substance in a solid form can penetrate to the interior of the fibre. The cloth, therefore, must be first impregnated with a liquid, by the decomposition of which the insoluble substance is to be produced.

The form in which a mordant exists on a piece of cloth ready to be dyed is usually that of a subsalt; that is, a body of a saline constitution (consisting of an acid and a base), in which the proportion of base is in considerable excess. When a piece of cotton, for example, is moistened with a solution of basic alum (soluble subsulphate of alumina) and dried, an insoluble subsulphate of alumina is produced on the cloth, containing less sulphuric acid than exists in the soluble subsulphate. It is not probable, however, that, when the cloth thus mordanted is immersed in the dye-beck, the insoluble sub-salt combines as such with the colouring matter in solution; the combination, which is doubtless a chemical one, takes place between the colouring matter and the base of the sub-salt. In this case, either all the acid of the sub-salt or else the corresponding soluble neutral salt is liberated and dissolved, the whole or the excess of base remaining on the cloth to fix the colouring material.

When a piece of cotton impregnated with subsulphate of alumina, for example, is put into the madder-beck, the colouring matter of the madder combines with either the whole of the alumina in the subsulphate, or else that portion of the alumina only which is in excess over the amount contained in the soluble sulphate. In the first case, the subsulphate of alumina is simply decomposed into alumina on the one hand, and sulphuric acid on the other; while in the second case it is resolved into alumina on one side, and the soluble and neutral sulphate on the other.

From the preceding observations may be inferred the necessity of distinguishing between the three states in which a mordant may exist; namely, first, in the soluble form in which it is applied to the cloth; second, as the insoluble sub-salt afterwards produced in the fibre; and third, as the true base existing in union with the colouring matter in the dyed cloth. The term mordant ought strictly to be confined to the true base, in whatever form it exists; but for convenience, it is also applied to the first solution, and to the sub-salt on the cloth before the dyeing.

In a few cases, however, the insoluble substance which is precipitated on the cloth from the solution of the mordant is not a sub-salt, but the true mordant or base itself; thus, pure alumina may be precipitated from the solution of aluminate of potash, and pure peroxide of tin from the solution of stannate of potash.

If the affinity of the colouring matter for the mordant is so powerful that a compound of the two is precipitated immediately on their solutions being placed in contact, the intermediate step, consisting in the formation of the insoluble sub-salt by drying, is sometimes omitted. The cloth is then first impregnated with the solution of the mordant, and after being washed, drained, and squeezed, is passed through the solution of the colouring matter. It frequently happens, however, and especially with wool and silk, that even in these cases an insoluble sub-salt does attach itself to the fibre from the solution; but it is commonly produced on cotton during the desiccation of the mordanted goods.

In all cases where the formation of an insoluble sub-salt by the desiccation of the mordant on the cloth is omitted, the excess of mordant which remains on the surface of the cloth is removed by washing and draining, before the cloth is exposed to the dyeing liquid. This excess of mordant is removed for three reasons; 1°, to prevent an unnecessary abstraction of colouring matter from the dye-beck; 2°, to ensure a more uniform distribution of tint; and 3°, to attach the colour in a more permanent manner. When the solution of the mordant produces an immediate precipitate with that of the colouring matter, if the excess of mordant is allowed to remain on the cloth, the latter assumes a good colour when put into the dye-beck, but which is, for the most part, merely attached to the surface, very little existing in the interior of the fibre. This seems to arise from the closing of the apertures through which a liquid obtains access to the interior of the tube or cell, by the precipitate of colouring matter.

But the colour which results when the washed and drained cloth is put into the dye-beck, is often devoid of lustre and very subject to alteration, apparently, because the quantity of mordant on the cloth is too small to form an intimate combination with all the colour which is deposited. On applying, however, either the same mordant a second time, or else another mordant, the brilliancy of the tint is greatly increased, and the colouring material becomes permanently attached. The second mordant which is applied in this manner is known among dyers by the name of *alterant*, proposed by Dr. Bancroft. A particular example will render such a process more intelligible. If a piece of white cotton cloth is transferred at once from a dilute solution of perchloride of tin to a weak decoction of logwood, the cloth assumes an uneven violet colour, feebly attached and removable by washing. But if the perchloride remaining on the surface is thoroughly removed before the cloth is put into the decoction, the piece assumes a dull brownish violet tint; by properly adjusting the strength of the solution of perchloride to that of the decoction, the latter may be entirely deprived of colour. If, in the next place, a small quantity of a solution of perchloride of tin, or acetate of alumina, is added to the liquor as an alterant, the cloth acquires a good violet or purple colour, and is now permanently dyed.

The most probable explanation of the action of the alterant in such a case as the above, is the following. With a proper proportion of the colouring matter of logwood, peroxide of tin forms a compound possessed of a lively colour, which compound is capable of uniting loosely with more of the colouring matter of logwood, the proper tint of which, by itself, is red. The colouring matter in excess does not partake of the lively violet or purple tint, of what may be considered as the neutral compound with the mordant, but the effect of applying more mordant (as the alterant) is obviously to form a neutral compound with the excess of colouring matter possessing the proper violet or purple colour. If the decoction of logwood to which the mordanted and washed cloth is exposed, is mixed with a very small quantity of a free acid, the precipitation of an excess of colouring matter is prevented (partly, it would seem, through the solvent power of the acid); so that the cloth assumes a lively colour at once, and the application of an alterant is unnecessary.

It is not essential to the character of an alterant that its action partake of that of a mordant. Thus, instead of mixing a free acid with the decoction of logwood, the acid may be afterwards applied to the dyed cloth, in which case it becomes the alterant, partly by removing some of the colouring matter in excess, and partly by disintegrating the particles of the mordant, whereby the latter is enabled to form a more intimate combination with the colouring matter.

In an extended sense, the term alterant may be applied to any substance which can effect a permanent change in the colour of a dyed cloth, whatever may be its chemical action. Thus, oxalic acid becomes an alterant when applied to the purple woollen cloth obtained by cochineal with a mordant of protoxide of tin, whereby the purple becomes changed to scarlet; bichromate of potash may also be called an alterant when applied to a piece of cotton dyed violet with logwood and alumina, in order to change the violet into a black.

In a few cases, where the affinity of the colouring matter for a mordant is not sufficiently strong for the former to separate, by itself, the mordant from the substance with which it is already in combination, the solution of the mordant and that of the colouring material may be mixed without the formation of a precipitate. Thus, no precipitate of colouring matter and mordant ensues when the soluble combination of alumina and caustic potash (aluminate of potash) is mixed with a solution of the colouring matter of annatto; nor when a solution of protochloride of tin is added to a *cold* decoction of logwood. When this is the case, the cloth is not always first impregnated with the mordant and afterwards with the colouring matter, but both may be applied at once by exposing the cloth to the mixture previously made. So far as the mere operation is concerned, however, this style of dyeing differs from that we have been considering, being referable to the third style according to the division followed in this section (page 262); but the function of the mordant and its action on the dye are of precisely the same nature as when the mordant is first applied, excepting that the formation of the insoluble compound of mordant and colouring principle which attaches itself to the fibre, appears here to be partly determined, as an induced effect, by the surface attraction of the tissue for the resulting compound. A similar effect may be produced in the above mixture of protochloride of tin and decoction of logwood by the application of heat, which seems to increase the mutual affinity of the mordant and colouring principle. The cloth does not in such cases abstract the whole of the colouring matter from the dyeing liquor, but the depth of its colour corresponds, in general, to the strength of the infusion, the temperature, and the time during which it is exposed to the mixed liquid.

Vegetable colouring matters are not often applied to goods of cotton by this style of dyeing except as steam colours, where the mixed solution of mordant and colouring matter is printed on the cloth, and the fixation afterwards effected by exposure to steam. The principal colouring principles thus applied are those of logwood, French berries, cochineal, peachwood, and quercitron, with a tin or aluminous mordant: an orange colour is sometimes imparted to cotton by means of the mixture of annatto and alumina in caustic potash, applied as above.

It rarely happens that the common aluminous mordants, such as basic alum and red liquor, can be employed alone as mordants for vegetable colouring matters when applied as steam or topical colours, on account of the facility with which the alumina is precipitated by the vegetable decoctions. But this tendency to precipitation may be greatly diminished by applying to the alumina a more powerful solvent. Thus, the solutions of alumina commonly employed with logwood, &c., when applied as steam colours, are basic alum mixed with oxalic or another strong acid, red liquor mixed with oxalic acid, and common alum mixed with sugar.

As the combination of the colouring principle with the mordant is to be considered a case of true chemical union, it might be anticipated that the colouring matter often experiences a considerable modification in tint on uniting with a mordant, independent of the colour possessed by the mordant itself, consistently with the law that a change of properties always attends chemical combination. Such a change in colour is certainly often exhibited, but sometimes the colour of the resulting compound is intermediate between that of the colouring matter and that of the mordant; and the colour of the dye-liquor may generally be imparted to a tissue without much alteration by the use of a white mordant.

If a piece of cloth is impregnated with alumina and then passed through the madder-beck, it acquires a rose tint; in the same dye-beck, but with peroxide of iron as the mordant, the cloth becomes dark brown or even black; and with a mixture of alumina and peroxide of iron, a puce colour may be obtained. If three pieces of cotton, one impregnated with alumina, another with peroxide of iron, and the other with oxide of copper, are passed through a decoction of quercitron, the piece containing alumina presents very nearly the proper yellow of the dye; that with oxide of iron has a dark fawn colour, and that with oxide of copper a yellowish fawn colour. Cochineal produces a purple compound with a mordant of protoxide of tin, which is itself white;

but a scarlet compound with peroxide of tin, which is also white.

The quantity of the colouring matter absorbed by the mordanted cloth is, in general, proportional to the quantity of mordant it contains, which may be determined by the strength of the solution of the mordant. Hence two pieces of cloth impregnated with solutions of the same mordant of unequal strength would present very different depths of colour by being passed through the same dye-beck. With the acetate of iron as a mordant, the same infusion of madder may be made to afford any variety of tint between faint lilac and black, and with an aluminous mordant any shade between the most delicate pink and dark red.

The class of bodies from which mordants are derived is the metallic oxides or bases, whence they might be supposed as numerous as metals which form insoluble oxides; but so many qualifications are required in a mordant, besides insolubility, that their number is very limited. Those most commonly employed are alumina, peroxide of iron, peroxide of tin, and protoxide of tin. The chemical affinity of these bases for colouring matters has already been alluded to (page 237); and many of their salts, or compounds with acids, have likewise a considerable attraction of surface for the stuffs, so that the latter have the power of withdrawing them to a certain extent from their solutions. These bases also possess a property which seems often to be highly advantageous, namely, that of forming with the same acid, when united with it in different proportions, a soluble and an insoluble combination.

Alumina,* which is the most extensively employed of the mordants, may be applied to cloths from four different solutions; 1°, from ordinary alum, which is a double salt composed of sulphate of alumina and sulphate of potash; 2°, from the solution of subsulphate of alumina, also known as basic alum, which is merely common alum with a portion of its acid neutralized by an alkali; 3°, from red liquor, which is commonly a solution in acetic acid of the subsulphate of alumina insoluble in water; and 4°, from a solution of alumina in caustic potash, known as the aluminate of potash. From alum, from the soluble subsulphate of alumina, and from common red liquor an insoluble subsulphate is produced on the cloth containing less sulphuric acid than the soluble subsulphate; and alumina itself is deposited from the aluminate of potash.

1. *Alum.*—This salt, in its ordinary state, is not much employed at present in cotton dyeing, being superseded by basic alum, red liquor, and the aluminate of potash; but it is still extensively used in silk dyeing and printing, and is the principal mordant used in the dyeing of wool. On cotton goods common alum is sometimes employed as the mordant for topical and steam colours, where the alu-

* This earth is thrown down as a white bulky precipitate when a solution of alum is mixed with an excess of ammonia. It is not obtained pure, however, by such a process, but retains some of the sulphuric acid. To prepare pure alumina, the precipitate as thus obtained may be redissolved in dilute sulphuric acid and again precipitated by ammonia. The proper neutral sulphate of alumina is very soluble in water and difficult to crystallize; by the addition of sulphate of potash it becomes common alum, which is much less soluble and very easily crystallized.

mina is required to be held in solution pretty strongly. The chemical constitution and properties of alum will form the subject of a future paper.

The impurity which common alum is likely to contain of greatest moment to the practical dyer is *iron*, existing in the state of sulphate (usually, both of the peroxide and protoxide), and chloride. It is derived from the mother-liquor from which the alum was crystallized.

The readiest method of detecting this impurity is by adding to the aqueous solution of alum a few drops of a solution of yellow or red prussiate of potash, which cause the formation of a precipitate of Prussian blue with a very small trace of iron. The precipitate at first produced with the yellow prussiate is generally greenish-white, but becomes blue on exposing the mixture to the air for a few minutes. In this case the iron is present in the alum in the state of a salt of the protoxide; and a solution of the red prussiate, instead of the yellow, causes the immediate production of Prussian blue. If the solution of alum to be tested is first boiled with a few drops of nitric acid to convert the protoxide of iron into peroxide, Prussian blue is immediately formed on applying the yellow prussiate.

In the few processes of cotton dyeing in which alum is employed as the mordant, the strength of the solution of alum and the manner of applying it entirely depend on the nature of the dye and the depth of the colour to be produced. In most cases, where the entire surface is to be impregnated with the colouring material, the goods may be allowed to remain for twenty-four hours, or more, in a cold solution containing about one part of alum to four parts of the cloth, with as much water as is requisite to cover the cloth. Silk may be digested for from eight to twenty-four hours, or longer, in a cold solution of three or four parts of alum to ten or twelve parts of silk and one hundred of water. Wool may be impregnated with alum by heating it in a solution containing of alum from one-sixth to one-fourth of the weight of the wool, and maintained at the boiling point for one hour, or a little longer; but ebullition for two hours is prejudicial. A very common addition to alum when employed as a mordant, especially for wool which is intended to be brightened by an acid after being dyed, is tartar (crude cream of tartar), which diminishes the tendency of alum to crystallize, and brightens the resulting colour. The quantity which is added varies for different dyestuffs from one-fourth to one-half of the weight of the alum. When employed as a mordant for topical or steam colours, a quantity of sugar is usually added to the alum.

2. *Basic alum.*—As common alum, of itself, has very little disposition to form an insoluble subsalt, it is a weak mordant for cotton goods, and has hence been superseded by other aluminous preparations, from which subsalts may be more easily produced. One of these is basic alum, which is made by separating from common alum a portion of its acid by the application of an alkaline carbonate. It is found that one-third of the acid contained in common alum is sufficient to retain in solution all the alumina of the alum, provided the liquid is cold and in a concen-

trated state. The partial separation of the acid may be effected either by carbonate of potash or carbonate of soda, added until it begins to produce a permanent precipitate. A gelatinous precipitate is formed from the first addition of the alkaline carbonate, but it redissolves on stirring until two-thirds of the quantity necessary for complete saturation has been applied.* In this state alum is a powerful mordant, as the excess of base is held in solution very feebly, and is easily removed in the state of an insoluble subsulphate through the surface attraction of the tissue. Animal charcoal also readily withdraws the excess of alumina, by virtue of the same force.

A solution of subsulphate of alumina is also produced when chalk is digested in a solution of common alum, sulphate of lime being then formed. According to Hausmann, one part of alum, with the addition of one-eighth part of chalk, may be retained in solution during summer by five parts of water. One part of common alum requires between eighteen and nineteen parts of cold water for its solution.

A solution of basic alum prepared by an alkaline carbonate, as above, cannot be well employed except in a concentrated state, as mere dilution with water determines the formation of a precipitate of insoluble subsulphate of alumina. But this inconvenience may be overcome, and an excellent mordant obtained, by adding acetic acid to the solution of basic alum. The mixture thus formed, which is

* A convenient mode of preparing the solution of basic alum is to dissolve in water two-thirds of the quantity of common alum operated on, to add carbonate of soda to the solution until the mixture exhibits a slight alkaline reaction, and then to add the remaining one-third of alum with agitation.

quite analogous to common red liquor, affords no precipitate on dilution with water.

3. *Red liquor and acetate of alumina.*—Red liquor is much more extensively employed as a mordant than any other preparation of alumina. The common method of preparing this liquid for the use of the dyer and calico-printer is by adding a solution of acetate of lead or acetate of lime to a solution of alum, when a portion of the sulphuric acid of the alum combines with the oxide of lead or the lime of the acetate to form an insoluble sulphate, and the acetic acid previously in combination with oxide of lead or lime combines with alumina to form a soluble acetate. To produce complete decomposition both of the sulphate of alumina and the sulphate of potash in the alum, with formation of sulphate of lead and acetates of alumina and potash,

478 parts, or 1 eq. of alum, require
756 parts, or 4 eqs. of crystallized acetate of lead.

And for the complete decomposition only of the sulphate of alumina in the alum,

478 parts, or 1 eq. of alum, require
567 parts, or 3 eqs. of crystallized acetate of lead.

The reactions which occur on mixing solutions of these materials in the latter proportions are expressed in the following diagram:

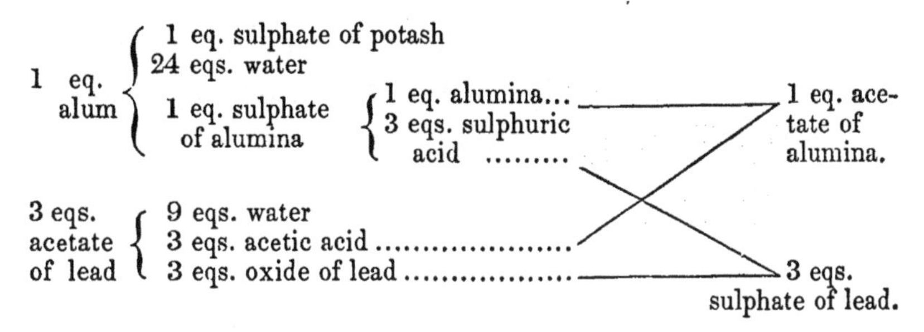

But the quantity of acetate of lead employed in the preparation of red liquor is never greater than that of the alum, and commonly one-third less, the proportions being slightly varied according to the purposes for which the mordant is required. A small quantity of carbonate of soda (from one-twentieth to one-tenth of the weight of the alum) is also sometimes added to the mixture to separate a portion of the sulphuric acid contained in the excess of alum.

The following proportions of the materials afford a strong mordant of specific gravity about 20° Twaddell* (1100), well adapted for producing dark reds with madder:

No. 1.

5 gallons of water,
10 pounds of alum,
1 pound of soda crystals,
10 pounds of acetate of lead.

The alum is first dissolved in boiling water, and to this solution the soda is added gradually; when the effervescence is subsided, the acetate of lead is added in a state of fine powder, and the mixture having been well agitated is allowed to stand for the sulphate of lead to settle, after which the supernatant liquid may be decanted for use.

* Degrees on Twaddell's hydrometer may be converted into the ordinary sp. gr. formula (water being 1000) by multiplying them by 5 and adding 1000.

A red liquor, better adapted than the above for producing a yellow dye with the colouring matter of quercitron, may be made by mixing, in the same manner,

No. 2.
5 gallons of water,
10 pounds of alum,
1 pound of soda,
$7\frac{1}{2}$ pounds of acetate of lead.

In consequence of the expense of acetate of lead, this salt is commonly superseded, in the preparation of red liquor, by acetate of lime, obtained by neutralizing with quick-lime the crude acetic acid or pyroligneous acid afforded by the distillation of wood; but the red liquor thus prepared does not produce with colouring matters such delicate and bright shades as that prepared by acetate of lead. The usual proportions of acetate of lime and alum employed for this purpose are two pounds and a half of the latter to a gallon of solution of the former of specific gravity 12° or 13° Twaddell. As met with in commerce, red liquor usually has a spec. grav. about 18° Twad.

The following mode of preparing red liquor by acetate of lime is recommended by M. Kœchlin-Schouch (*Bulletin de la Société industrielle de Mulhausen*, t. i. p. 277). In twenty-five gallons of hot water dissolve two hundred pounds of alum, and to the solution add three hundred pounds of the crude solution of acetate of lime of specific gravity 16° Twad. The resulting red liquor has the density, while hot, of 22° Twad., but on cooling it deposits crystals of alum, and falls in specific gravity to 18° Twad.

In neither of the preceding preparations is sufficient acetate of lead or acetate of lime employed to decompose the whole of the sulphate of alumina in the alum, and it is doubtful, moreover, whether acetate of lime, in any quantity, would effect the complete decomposition of sulphate of alumina. But this undecomposed alum or sulphate of alumina, instead of being useless, as some have supposed, forms a highly important constituent of the mixture. By its action on the acetate of alumina in the solution, it gives rise to the formation of subsulphate of alumina or basic alum, and free acetic acid, and the latter serves to retain the former in a state of more permanent solution than water would alone.

On applying heat to red liquor, a precipitate of subsulphate of alumina is produced in the liquid, containing, according to the analysis of M. Kœchlin-Schouch, eight equivalents of alumina and three equivalents of sulphuric acid, or, eight times as much alumina as the neutral sulphate in common alum. The temperature at which the precipitation commences varies according to the strength of the liquor and the proportions of acetate of lead and alum employed in its preparation. When made as No. 1, page 281, the precipitation commences at about 154° Fahr. If the source of heat is withdrawn soon after the precipitate appears, so as to avoid the evaporation of acetic acid and the aggregation of the precipitate, the latter completely redissolves as the liquid cools; but if the heating is continued until a sensible quantity of the acetic acid is evaporated and the precipitate is become dense, the subsulphate does not redissolve on cooling, nor even on the addition of free acetic acid. Such a precipitation of insoluble subsulphate, accompanied with the evaporation of acetic acid, always occurs during the drying and "ageing" of cottons printed with red liquor.*

A solution of pure acetate of alumina obtained by dissolving recently precipitated hydrate of alumina in acetic acid is uncrystallizable, and dries, on evaporation, into a gummy mass, very soluble in water. The aqueous solution of the pure acetate may be boiled without decomposition; but if a solution of alum is added to acetate of alumina, so as to form a mixture analogous to red liquor, the liquid affords, on the application of heat, a precipitate of subsulphate, of the same composition as that produced from common red liquor, which redissolves on the cooling of the liquid if the acetic acid has not been expelled.

Acetate of alumina made without excess of alum is very rarely used as a mordant, the proportions of alum and acetate of lead employed in almost all cases being four parts of the former to three parts of the latter. The chief use of the pure acetate, or rather of the mixture of pure acetate with sulphate of potash, such as is obtained by mixing eight parts of alum with nine and a half parts of acetate of lead, is to add to mixtures for topical colours containing a strong acid, such as muriatic, sulphuric, or nitric, in the free state. The strong acid combines with the

* Concentrated red liquor deposits a small quantity of the subsulphate of alumina at common temperatures, if kept for a considerable time. The precipitate thus gradually formed is sometimes too aggregated to be redissolved on the application of acetic acid.

alumina of the acetate, and liberates acetic acid, which exerts no corrosive action on the fibre of the cloth.

4. *Aluminate of potash.*—Another preparation of alumina much employed as a mordant for cotton goods, is the solution of alumina in caustic potash, known as the aluminate of potash. The following method of preparing this solution is recommended by M. Kœchlin-Schouch. A solution of caustic potash is first made by boiling for half an hour a mixture of eighty pounds of carbonate of potash, thirty-two pounds of quick-lime, and forty gallons of water. The caustic ley being allowed to settle, thirty gallons are decanted and evaporated down to the density 35° Baumé (60° Twaddell), and sixty pounds of powdered alum are added to the boiling liquid. As the solution cools, a quantity of sulphate of potash is deposited in crystals.

When a piece of cloth impregnated with the aluminate of potash is suspended freely in the air, the carbonic acid of the atmosphere seizes upon the caustic potash which holds the alumina in solution, causing the formation of carbonate of potash and precipitation of alumina. If the apartment in which cottons printed with the aluminate of potash are suspended is imperfectly ventilated, after a short time not a trace of carbonic acid can be detected in the atmosphere by the ordinary test of lime-water; hence the necessity of paying particular attention to the means of producing a proper ventilation in the "hanging" or "ageing" room, if the complete precipitation of the alumina during that stage of the process is required.

The time of hanging the mordanted goods, however, is seldom prolonged sufficiently to allow of the complete decomposition of the aluminate of potash. This is ensured by afterwards passing the cloth through a dilute solution of muriate of ammonia,* which immediately determines the complete precipitation of the alumina. The reactions which take place when a solution of aluminate of potash is mixed with a solution of muriate of ammonia are expressed in the following diagram:

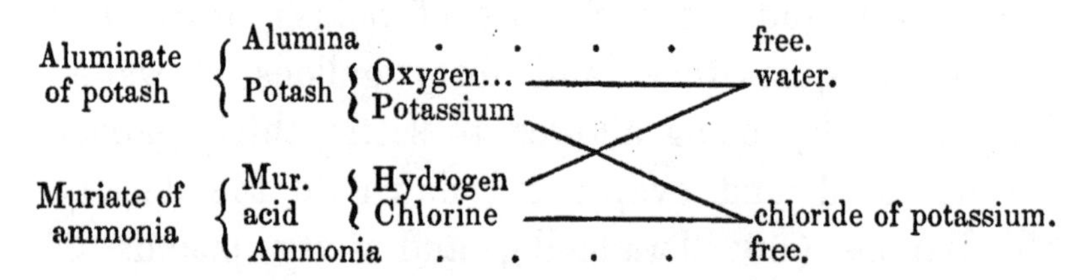

The aluminate of soda may be prepared in the same manner and used for the same purposes as aluminate of potash. It is said that no difference is perceptible between the effects obtained by aluminate of potash, and those by aluminate of soda.

The other simple preparations of alumina which are occasionally used as mordants are, nitrate of alumina, chloride of aluminum, and tartrate of alumina. Of these, the most extensively employed is the nitrate, which may be prepared of sufficient purity for the use of the dyer and calico-printer by mixing concentrated solutions of equal weights of alum and nitrate of lead, when sulphate of lead is formed and

* The muriate of ammonia is sometimes mixed with the dung-beck, and sometimes with the solution of "dung substitute." In a few particular styles of calico-printing, where the ageing of cottons printed with aluminate of potash is altogether omitted, the cloth is passed through a solution of muriate of ammonia before the dung emulsion.

precipitated, and nitrate of alumina remains in solution.

Tin mordants.—Several preparations of tin are employed as mordants in dyeing and calico-printing, comprising salts of the protoxide and of the peroxide, and mixtures of the salts of both oxides. The oxides of tin have a strong tendency to unite with soluble vegetable and animal colouring matters, producing distinct and definite combinations; and the compounds with the peroxide are generally distinguished for possessing a vivacity of tint far superior to that presented by the combinations of the same colouring matter with any other mordant.

Peroxide of tin is used as a mordant chiefly with cochineal, Brazil-wood, peachwood, barwood, French berries, and logwood, and is commonly applied in the state of a solution of the perchloride (permuriate), or as a mixture of the solution of the perchloride with that of the pernitrate. Such solutions, which are known among dyers by the name of *red spirits* or simply *spirits*, may be obtained by dissolving metallic tin, in a granulated or "feathered" state, in one of the following liquids:

1°. Aqua regia, which is a mixture of nitric and muriatic acids;

2°. A mixture of nitric acid and muriate of ammonia; and

3°. A mixture of nitric acid, muriate of ammonia, and common salt.

The perchloride of tin, or a mixture of the perchloride and pernitrate, is also sometimes prepared from crystals of the protochloride (salts of tin) by means of nitric acid or aqua regia. The nitric acid used for this purpose should be quite free from sulphuric acid.

A great number of receipts for the preparation of this mordant have been prescribed, varying very considerably in the proportions of the materials, according to the nature of the fabric to be dyed, and that of the dye-stuff for which it is to be used as the mordant. Some of the preparations contain the peroxide or perchloride only; but others, which are preferred for general use, contain both the perchloride and the protochloride. A common process for preparing a mixture of the two chlorides is to add granulated tin very gradually to a mixture of three parts by measure of muriatic acid, and one part of commercial nitric acid, so long as any tin is dissolved in the cold. If the tin is not added gradually, instead of being dissolved, it is converted into the insoluble peroxide, which is deposited as a white powder.

The above proportions answer well for a mordant for general use, and especially for Brazil-wood; but for particular purposes the proportions of muriatic and nitric acids are varied from six parts of the former, and one of the latter, to equal parts.

The solution of the perchloride of tin, or mixed perchloride and protochloride made by dissolving tin in a mixture of nitric acid and sal-ammoniac, is much used by silk and woollen dyers, but a considerable difference exists between the proportions of the materials as recommended by different dyers. For general purposes, the solution afforded by the following proportions receives a decided preference:

3 quarts of nitric acid of specific gravity 1·300,
4 quarts of water,
12 ounces of muriate of ammonia,
30 ounces of granulated tin.

The muriate of ammonia is first dissolved in the mixture of acid and water, and to this solution the tin is added in small quantities at a time, so as to prevent the mixture from becoming very hot.

The salt met with in commerce under the name of *pink salt* is the double perchloride of tin and muriate of ammonia (chloride of tin and ammonium), which is made by adding muriate of ammonia to a solution of the perchloride, and evaporating to obtain crystals. It is chiefly used as a mordant with peachwood.

Peroxide of tin is often applied to cloth in the state of the soluble combination of caustic potash and oxide of tin, known as stannate of potash, which may be obtained by adding a solution of caustic potash to a solution of perchloride of tin, until the precipitate at first produced is entirely redissolved. If a piece of cotton impregnated with such a solution is dipped into dilute sulphuric acid, or a solution of muriate of ammonia, the alkaline combination on the cloth is decomposed, and peroxide of tin precipitated within the fibre. The decomposition which ensues on mixing stannate of potash with muriate of ammonia is quite analogous to that which occurs on the mixture of aluminate of potash with muriate of ammonia (page 286).

Protoxide of tin is frequently used as a mordant

alone, as well as the peroxide. This oxide may be applied from the protochloride of tin, which is prepared by dissolving metallic tin in pure muriatic acid to saturation, with the assistance of heat. One part of tin may be dissolved in about three parts of concentrated muriatic acid, and on evaporation the solution affords small colourless crystals, distinguished as *salts of tin*. The solution of the protochloride is known among dyers by the name of *plum spirits*, being used in the preparation of the *plum tub*, which is a mixture of decoction of logwood with the protochloride.

This salt has several interesting applications in calico-printing, both as a mordant and a deoxidizing agent, to which we shall again have occasion to advert. The solution of protoxide of tin in a caustic alkali, obtained by adding the alkali to the solution of protochloridc of tin until the protoxide at first precipitated is redissolved, is frequently used in the place of the protochloride.

When exposed to the air, a solution of protochloride of tin absorbs oxygen, and affords, if not very acid, a white precipitate consisting of a subsalt of the peroxide. This inconvenience may be counteracted to a great extent by the addition of muriate of ammonia which combines with the protochloride to form a double salt, less disposed to absorb oxygen than the pure protochloride.

The colours of the compounds of colouring matters with peroxide of tin are generally much brighter than those of the same compounds with protoxide of tin, but solutions of the protoxide enter the pores of cotton fabrics better than solutions of the peroxide.

On this account, a practice sometimes pursued in dyeing cotton goods by a tin mordant, is first to apply the tin in the state of protochloride, and to form the peroxide afterwards, within the fibre, by wincing the goods in a dilute solution of chloride of lime.

Iron mordants.—The principal simple preparations of iron which are employed as mordants are the following: copperas, which is the sulphate of the protoxide; iron liquor, which is an impure acetate of the protoxide; the pernitrate, the sub-persulphate, and the perchloride. The most available of these forms of iron is copperas; but this salt is not well adapted as a mordant for cotton goods, as the powerful affinity of sulphuric acid for protoxide of iron is an impediment to the formation of an insoluble subsalt.

Acetate of iron; iron liquor.—The iron mordant commonly used in calico-printing is the acetate, which may be prepared by mixing a solution of acetate of lime or acetate of lead with a solution of copperas. A double decomposition occurs on the mixture of these solutions, with the formation of sulphate of lime or sulphate of lead, which falls as a heavy precipitate, and acetate of protoxide of iron, which remains in solution. For the complete decomposition of copperas by acetate of lead, 10 parts of the former require about 13½ parts of the latter; but in the preparation of acetate of iron in this way on the large scale, the copperas is always employed in excess, being seldom in so small a proportion to the acetate of lead as an equal weight. By exposure to the air the acetate of the protoxide becomes partially peroxidized, being converted into a subacetate of the peroxide.

But nearly all the acetate of iron used in print-works is now prepared by digesting, for several weeks, old iron hoops, nails, &c., in the crude acetic acid obtained by the distillation of wood. A dark brown solution, known as the pyrolignite of iron or iron liquor, is thus obtained, composed of the acetate of the protoxide of iron, and a quantity of tarry, oily, and spirituous matters produced in the destructive distillation of wood. As a mordant, this mixture is in general preferred to the purer article prepared by means of acetate of lead or acetate of lime, probably because the peroxidation of the protoxide of iron by exposure to the air during the "ageing" of the goods is retarded by the spirituous and unctuous matters present which have a stronger affinity for the oxygen of the air. A small quantity of the acetate of the peroxide of iron is sometimes contained in iron liquor, but by no means as an essential constituent.

The principal pure persalt of iron used in dyeing and calico-printing is the nitrate, which is prepared by dissolving clean pieces of iron in nitric acid of specific gravity 1·305. Soon after the evolution of brown fumes ceases, the acid solution should be decanted, so as to avoid the formation of the insoluble sub-pernitrate of iron. This solution of iron is used as a mordant with vegetable colouring matters, and also for producing a buff colour with an alkali (see page 256), and Prussian blue with yellow prussiate of potash.

A preparation of iron extensively employed at some print-works in the place of the common acid pernitrate, is a mixture of the neutral pernitrate with free acetic acid, obtained by adding about a pound of powdered acetate of lead to two pints of a solution of the pernitrate, of the density 1·55. The acetate of lead is decomposed by the free nitric acid present in the solution, with formation of nitrate of lead, which is precipitated, and free acetic acid.

A solution of a sub-pernitrate of iron, made by adding a small quantity of an alkaline carbonate to the pernitrate, is also sometimes advantageously substituted for the pernitrate prepared as above. The peroxide of iron at first precipitated may be redissolved on agitation, if only a small proportion of alkali has been applied.

Two other forms of peroxide of iron have been occasionally employed as mordants; one analogous in its chemical constitution to basic alum (page 278), and the other to red liquor (page 280). The first is prepared by partially decomposing, by means of an alkaline carbonate, the persulphate of iron made by boiling copperas in dilute nitric acid. The oxide at first precipitated by the alkali is slowly redissolved by the undecomposed persulphate, giving rise to the subsulphate of the peroxide. The preparation of peroxide of iron, analogous to red liquor, may be made by adding one part, by weight, of acetate of lead to four parts of a solution of persulphate of iron of the density 1·65. Sulphate of lead is precipitated, and the solution comes to contain subsulphate of the peroxide of iron, and peracetate of iron or free acetic acid.

A patent has been recently obtained by Mr. Mercer and Mr. Barnes for the preparation of an acidulous liquid for mixing with ferruginous mordants, which imparts considerable brilliancy to the tints of all compounds of colouring matters (particularly that of madder) with peroxide of iron. This liquid, known as "assistant mordant," or "patent purple liquor," is made by digesting farinaceous substances, starch, or sugar, in warm nitric acid, of moderate strength. The temperature of the mixture is not allowed to become high, in order to prevent the formation of oxalic acid. When all the nitric acid present is decomposed, a little pyroligneous acid is added, and the mixture is then ready for use.

For producing a black dye, with oxide of iron and madder, one measure of iron liquor may be mixed with one or two measures of "assistant;" for light purples, the proportion of "assistant" may be increased to five or six measures. The beneficial action of this preparation is considered to consist chiefly in the retardation of the peroxidizement of the iron mordant during the "ageing."

The enlivening action of strong acids and acidulous salts on the tints of some vegetable and animal colouring matters, when applied in a diluted state to the compounds of colouring matters and mordants, is a subject which deserves the particular attention of the dyer and calico-printer. Instances of this effect of the application of acids, either to the dyed cloth or to the infusion of the colouring matter, are of frequent occurrence; and some bright shades of colour cannot be obtained without the assistance of an acid, though the mode of action of the acid is by no means obvious.

If a piece of cotton is printed or padded with iron liquor, dunged, winced in a mixture of chalk and hot water, and then dyed in a decoction of *valonia* or other vegetable matter containing tannin, it acquires merely a dull drab colour. But if the cloth thus dyed is exposed for a short time to the fumes of hydrochloric acid, or if winced in very dilute sulphuric or hydrochloric acid, the colour of the cloth changes from drab to slate colour or black, according to the quantity of mordant on the cloth. Acetic acid produces the same effect, but not in so powerful a manner as sulphuric acid, of which a mere trace is as effective as a much larger quantity. The colour of the compounds of both oxides of tin (but especially that of the protoxide) with the colouring matter of logwood may also be greatly enlivened by the application of very dilute sulphuric acid, and a good scarlet or crimson cannot be obtained from cochineal with a tin or aluminous mordant without the introduction of tartaric or oxalic acid or cream of tartar.

That the mode of action of the acid in such cases varies with different colouring matters and mordants hardly admits of doubt. The quantity of acid required in some cases is far too small to warrant the supposition that it enters into a permanent chemical union with the colouring matter and mordant; but in other cases a sensible quantity of acid disappears, and seems to enter as an essential constituent into the composition of the colouring material.

In cases where the acid does not enter into a permanent chemical combination with the mordant and colouring matter, it may be considered to act in two ways:

1. By preventing the deposition of an excess of the colouring matter on the mordanted cloth, or by removing such an excess, if already deposited, through its solvent power;

2. By effecting the disintegration of the particles of the mordant, whereby the latter is enabled to form a more intimate combination with the colouring material.

When a piece of cotton cloth impregnated with a mordant in the ordinary manner is immersed in the dyeing decoction, it often absorbs more colouring matter than is necessary to form what may be considered as the neutral compound of colouring matter and mordant. But this excess of the colouring principle is frequently prejudicial to the tint of the compound of the mordant with a smaller proportion of colouring matter, with which the excess is in a state of loose combination, being held, probably, by a mere attraction of surface (see page 237), and not by chemical affinity. Thus the compounds of both oxides of tin with an excess of the colouring matter of logwood are considerably inferior in vivacity of colour to the compounds of the same oxides with a smaller proportion of the colouring matter. The dullness of the colour of a piece of cotton mordanted with iron liquor, dunged, winced in hot chalky water, and dyed in a decoction of valonia (see page 295), may be attributed to the deposition of an excess of tannic acid (which is the astringent principle of the valonia) on the neutral and darker coloured tannate of iron.

Now one effect of the application of a small quantity of a strong acid to the decoction of the dye-stuff, may be the prevention of the attachment of this excess of colouring matter to the mordanted cloth; an acidulous liquid being generally a more powerful solvent of the colouring principle than pure water; and a corresponding effect may take place when the dyed cloth containing an excess of colouring matter is exposed to an acidulous liquid, the excess being then partly removed from the cloth, through the solvent power of the acid.

But a more important effect to be attained by the application of an acid is the disintegration of the mordant on the cloth, whereby the interior particles of the mordant become placed in a better condition for combining with and retaining the colouring matter. During the drying of a mordanted piece of cotton, the mordant always becomes more or less aggregated; and if the cloth is exposed to a high temperature, the aggregation may become so great with some mordants that very little colouring matter is absorbed when the cloth is immersed into the dyeing liquid containing no free acid. A remarkable difference may be observed between the quantity of the colouring matter of logwood absorbed by a piece of cotton impregnated with either perchloride or protochloride of tin, and dried at a very gentle heat, and the quantity of the same colouring matter absorbed by a piece of cloth containing the same mordant, but dried at as high a temperature as the cloth will well support. Even two pieces of cotton containing similar quantities of the same tin mordant, both dried at a low heat, but one washed in cold water after being dried, and the other in hot water, present a very sensible difference in shade when dyed in the same infusion of logwood. It is difficult to conceive how such effects are produced, except through differences in the state of aggregation of the mordant on the cloth.

But the addition of a small quantity of sulphuric acid to the dyeing liquid, causes the production of an uniform and bright colour in both pieces of cotton, whether dried at a high or low heat, or washed in hot or cold water. A mere trace of acid is generally sufficient to produce this effect, hence it cannot be supposed to act in all cases by forming a permanent chemical combination with the colouring material.

The disintegration of the precipitated mordant on the cloth seems to be the principal effect of the application of an acid in such cases as those now under consideration. The interior, uncombined, particles of the mordant thereby become exposed, and placed in a condition to unite intimately with the colouring material, the latter being derived either from the dyeing liquor or from the excess in loose combination with the exterior particles of the mordant. The disintegration of the mordant by the acid may possibly be a mere mechanical effect,* independent of any chemical alteration of the mordant, but the action of the acid admits of a more satisfactory explanation. The precipitated mordant is probably first dissolved by the acid, but is immediately reprecipitated in intimate combination with the colouring matter; the acid thereupon becomes liberated, and enabled to act in the same manner on other portions of the mordant. Whether the acid be mixed with the dyeing infusion, or applied as an alterant (see page 270) to the dyed cloth containing an excess of colouring matter, its action is of the same nature. An illustration of this principle is afforded by an experiment before referred to, which consists in exposing to the diluted fumes of muriatic acid a piece of cotton, with oxide of iron as the mordant, dyed to a drab colour in infusion of valonia (see page 295). The tannin or the astringent principle of the valonia is united only with the exterior particles of the oxide of iron, but is there contained in excess; on the application of the acid both the excess of tannin and the interior particles are dissolved, brought into contact, and precipitated in intimate combination, the muriatic acid being liberated to act on other portions of oxide of iron and tannin in a similar manner.

The preceding explanation of the enlivening ef-

* Examples of the disintegration of precipitates, by being digested in an acid liquid, which exerts no chemical action on the precipitates, are by no means rare. Thus, an intimate mixture of pure Prussian blue with hydrochloric acid appears to become a perfect solution on standing; but the particles of the pigment are merely disintegrated, and still exist in an insoluble form.

fects produced by the application of acids to compounds of organic colouring matters and mordants, refer only to those cases in which the quantity of acid sufficient to produce the effect is too small to warrant the supposition that it enters into a permanent chemical combination with the colouring material. In other cases a sensible quantity of free acid unites with the compound of mordant and colouring matter, and modifies the properties of the latter by causing a new disposition of its particles.

That the purple compound of the colouring matter of cochineal with protoxide of tin, for example, is capable of uniting with free sulphuric acid appears evident from the following experiment, communicated to me by Mr. Mercer. When a solution of protochloride of tin is added to a decoction of cochineal, a purple lake is precipitated, consisting of a compound of the colouring matter of cochineal with protoxide of tin mixed with an excess of the protoxide. This lake is collected on a filter, carefully washed with distilled water, and digested in a given quantity of a solution of carbonate of soda of known strength. The proper compound of colouring matter and oxide of tin then dissolves in the alkaline carbonate, and the excess of oxide of tin remains undissolved. On mixing an excess of sulphuric acid with the alkaline solution the pure lake is reprecipitated of a much richer colour than before being thus treated; but the reprecipitation of the pure lake requires considerably more acid than is necessary to neutralize the alkali. None of the lake is thrown down if no more acid is applied than the quantity exactly necessary for the neutralization; but on adding more, the precipitation immediately commences, though no free acid can be detected in the mixture until the lake is completely thrown down.

It is difficult to determine, precisely, the nature of the modification which colouring matters, or compounds of colouring matters and mordants, experience by combining with an acid in such a case as the preceding; some considerations on this subject, however, may not be without a practical application in the operations of the dye-house.

Soluble vegetable and animal colouring matters have the property of uniting both with acids and bases: with acids they form feeble combinations possessed of lively tints; but with bases they afford stronger combinations, having less vivacity of colour. In general, the more closely a mordant resembles an acid in its chemical character, the brighter are the tints of its compounds with colouring materials. Thus, peroxide of iron and peroxide of tin, which appear to act the part, sometimes, of feeble acids, produce compounds with colouring matters having brighter colours than similar compounds with protoxide of iron and protoxide of tin. But metallic protoxides, which possess a greater basic power than their corresponding peroxides, generally form more intimate combinations with soluble organic colouring materials, than the peroxides, especially if the protoxide is of itself a stable substance, as protoxide of tin. Some metallic protoxides also unite with a larger quantity of colouring matter than peroxides, but it is uncertain whether the excess is held by chemical affinity or by a mere attraction of aggregation dependent on the state of the surface of what we may consider as the neutral compound of colouring matter and mordant.

For tissues which can combine with and retain strong acids without injury to the fibre, as wool and silk,* protoxide of tin is as suitable a mordant as the peroxide, because the brightening effect which the mordant fails to produce may be obtained by the application of an acid. For this purpose oxalic acid is frequently applied to woollen goods containing protoxide of tin as the mordant.

But as very few strong acids can be applied to cotton goods without weakening the fibre to a greater or less extent, it becomes necessary, in order to obtain bright colours, to apply such mordants only as possess some resemblance to acids, as alumina, and the peroxides of iron and tin. In the action of such bodies on soluble organic colouring materials two forces may be recognized: by one, the mordant acts as an acid, producing an enlivening of tint; by the other, it acts as a base, producing an intimate and stable combination.

* If silk and wool are digested in dilute sulphuric or muriatic acid, a portion of the acid combines with the stuffs, and the liquid is found weaker after than before the immersion. The animal tissues also combine with tartaric acid when digested in a solution of cream of tartar (bitartrate of potash), leaving neutral tartrate of potash behind. But cotton exhibits no such disposition to unite with acids, and when digested in dilute sulphuric or muriatic acid, abstracts the water in preference to the acid, thus making the liquid more acid than before the immersion of the cotton.

DUNGING.

As the precipitation of the mordant in the form of an insoluble subsalt during the hanging or "ageing" of cotton goods is never complete, it becomes necessary to remove the unprecipitated mordant from the cloth before the dyeing, else a quantity of superfluous colouring matter will be deposited on the surface of the cloth, which would have to be removed by subsequent operations, besides causing an unnecessary impoverishment of the dyeing liquid. But the necessity for removing the superfluous mordant is chiefly experienced with cotton goods on which the mordant is printed so as to produce a pattern. If all the mordant which remains in a soluble form is not completely removed from such goods, a portion of it may become distributed over the whole surface of the cloth when the pieces are washed in water or put into the dye-beck.

One process for effecting the complete removal of the unprecipitated mordant consists in simply drawing the dried goods through a warm emulsion of cow-dung and water. The emulsion is usually contained in two stone cisterns, each about six feet long, by three feet wide and four feet deep: that in one cistern contains about two gallons of dung, to the cistern-full of hot water; that in the second contains only half this proportion of dung. The cloth, on being taken from the "ageing" room, is first drawn pretty quickly through the emulsion containing most dung, and immediately afterwards through the other; the cisterns being usually placed end to end, to allow the cloth to be conducted directly from the first to the second. The cloth is guided in its passage through the cisterns by four or five rollers placed at each end, it being essential to the success of this operation that the pieces should be extended and free from folds. Immediately on issuing from the second cistern, the cloth is passed over the reel of a contiguous wince-pit (see figs. 31, 32, page 314), where it is well washed in clean water, and from the wince-pit it is generally taken to the dash-wheel. If the mordant on the cloth is the aluminate of potash, some muriate of ammonia is added to the dung emulsion, to ensure the precipitation of alumina, which it effects by a reaction explained at page 286. Occasionally, the cloth containing aluminate of potash is passed through a solution of muriate of ammonia before it is exposed to the dung emulsion.

In calico-printing, the dunging process is necessary for all kinds of aluminous, iron, and tin mordants, when applied to the cloth before the colouring matter. The time of immersion, the temperature of the mixture, and the number of pieces which may be passed through a given quantity of dung and water, depend entirely on the state and quality of the mordants, and on the nature of the thickening paste by which the mordants are applied. A piece of cotton with a mordant which has a strong acid requires a longer time than a piece the mordant on which has a weak acid; and when the thickening paste for a mordant is flour or starch, a higher temperature is required than when British gum or common gum is used. The usual temperature of the dung emulsion is 160° or 180° Fahr.

Dunging is one of the most important steps in the process of calico-printing; and if badly performed, especially when the mordant on the cloth is alumina, the success of the subsequent dyeing is sometimes greatly endangered. The operation has for its object, not merely the removal of the superfluous mordant, and in printed goods of the thickening paste by which the mordant is applied, but the determination of a more intimate union between the mordant and the stuff, which it seems to effect by converting the sparingly soluble subsalts on the fibre into other compounds, perfectly insoluble.

Although the objects of the operation of dunging are sufficiently obvious, yet the precise manner in which they are attained is involved in some uncertainty. According to an analysis by M. Penot, cow-dung contains the following ingredients in 100 parts.

COMPOSITION OF COW-DUNG.

Woody fibre	26·39
Albumen	0·63
Chlorophyl	0·28
A sweet substance	0·93
A bitter matter	0·74
Chloride of sodium	0·08
Sulphate of potash	0·05
Sulphate of lime	0·25
Carbonate of lime	0·24
Phosphate of lime	0·46
Carbonate of iron	0·09
Silica	0·14
Water	69·58
(Loss	0·14)
	100·00

It is generally admitted that the superfluous or unprecipitated mordant is immediately dissolved by the hot water; but instead of remaining in a state of solution it is entirely precipitated in an insoluble form, partly by the albuminous constituent of the dung, partly by the phosphate and carbonate of lime, and partly by the insoluble ligneous fibre, and is therefore rendered incapable of attaching itself permanently to the cloth. But it appears that a small portion of the superfluous mordant dissolved from the cloth by the hot water, instead of being afterwards precipitated, is permanently retained in solution in a peculiar state of combination with the animal matter of the dung; one of the characters of which combination is, that it is incapable of affording a precipitate of subsalt or oxide to the cloth. The precise nature of this compound of the mordant with organic matter is not known, but it is believed to be analogous to that of several soluble combinations of metallic oxides with organic matters which are not affected by certain chemical reagents in the same manner as ordinary salts of such oxides. Thus, a salt of the protoxide of copper in the presence of several organic substances, sugar for instance, is not precipitated by a solution of a caustic alkali; the double tartrate of the peroxide of iron and potash does not afford a precipitate of peroxide of iron when mixed with a solution of caustic potash, nor does it yield a subsalt to cotton, like most other ferruginous salts.

The constituents of the dung which appear to be principally concerned in the fixation of the mordant on the cloth are the albuminous and soluble vegetable matters and the phosphate of lime. The former act partly by uniting with the base of the subsalt on the cloth, liberating its acid and forming a new combination more insoluble than the previous subsalt. The liberated acid may soon be detected in the liquid by the test of blue litmus paper, and requires to be neutralized, in a few processes, by the introduction of chalk.* The albuminous matter of the dung seems also to exercise an influence as a detergent or an emollient, whereby it considerably facilitates the detachment of the loosely combined mordant.

The action of the phosphate of lime in the dung emulsion is to cause the formation on the cloth of phosphate of alumina or phosphate of iron by a double decomposition, the acid in the subsalt uniting at the same time with the lime of the phosphate. Both phosphate of iron and phosphate of alumina are quite insoluble in water, and also in acetic acid, if warm.

Within the last few years the dung emulsion has been superseded, either partially or entirely, in all well conducted print-works in this country, by a solution of phosphate of soda and phosphate of lime, known by the name of "dung substitute," or simply "substitute," for the preparation of which a patent

* On the Continent, it is a common practice to add to the dung emulsion, in all cases, either chalk or bicarbonate of soda, the latter being preferred; but it is unusual in this country to make any such addition to the dung, except in cases where the cloth contains a free acid or acidulous salt, as lemon-juice or bisulphate of potash. The smallest excess of an alkaline carbonate should be avoided when alumina is the mordant on the cloth.

has been obtained by Mr. Mercer, Mr. Prince of Lowell, Massachusets, and Mr. Blyth. A solution of an alkaline arseniate had been long previously used as a substitute for dung by Mr. Mercer.

Dung substitute is prepared by mixing sulphuric acid with bone-earth, which consists chiefly of phosphate of lime; the acid not being applied in sufficient quantity to decompose the phosphate of lime entirely, but to produce an acid phosphate, or a solution of the phosphate in free phosphoric acid. Carbonate of soda is then added to neutralize the free acid completely, and the mixture is evaporated until the residuary mass becomes almost dry. When the concrete thus obtained is mixed with water, it affords a solution of phosphate of soda containing some phosphate of lime; a white mud remains undissolved consisting of sulphate, carbonate, and a little phosphate of lime, which should be carefully stirred up when the liquid is about to be used.

This preparation is not, of itself, an efficient substitute for all the essential, or at least, for all the important, constituents of the dung emulsion. To supply an emollient and detergent substance in the place of the albuminous matter of the dung, it is found necessary to mix with the above liquid a solution of glue or some other form of gelatine. The material employed for this purpose in most print-works is a solution of bone-size, called "cleansing liquor," which is made by boiling bones in water for nearly a week, separating the fat which rises to the surface of the liquid, and evaporating the aqueous solution of gelatine until it attains a density about 36° Twaddell (1·180). The advantage of making this addition to the phosphates was first pointed out by Mr. Mercer.

When the "aged" cloth is passed through a mixed solution of substitute and gelatine, the latter greatly facilitates the separation of the loosely combined mordant, and prevents its reattachment, while the phosphate of lime and phosphate of soda in the former serve to fix the alumina and oxide of iron in more intimate combination with the stuffs by converting them into phosphates. The acids previously in combination with the alumina and oxide of iron (when these bases existed as subsalts) unite at the same time with the soda and lime of the substitute.

The following detailed account of the best mode of applying the solution of substitute to mordanted goods has been communicated to me by Mr. Mercer. It refers to cases in which dung is entirely dispensed with, and in which the mordant is applied to the cloth topically.

The cloth is exposed to the action of two solutions of the substitute consecutively; that first applied, which is considerably stronger than the other, may be contained in a common dung cistern capable of holding not less than six hundred gallons, and furnished with a series of rollers so as to allow the immersion of fifteen yards of cloth at the same time. The weaker solution of substitute is applied to the cloth in a wince-pit.

A normal solution of substitute, called "substitute liquor," is first made by dissolving the substitute in warm water at the rate of two pounds to the gallon. Six gallons of this substitute liquor and two gallons of the cleansing liquor are introduced into the cistern, which is then filled with hot water, and the pieces of cloth are passed through at the rate of thirty yards per minute. The temperature of the solution may be the same, in general, as that of the dung-beck, in the common dunging process: for madder purples and pale reds it should never exceed 140° Fahr., but for madder blacks and full reds it may be a little higher. This cistern requires to be frequently renewed by the addition of fresh quantities both of substitute liquor and cleansing liquor. A gallon of the former and a quart of the latter may be added for every thirty or fifty pieces, according to the "heaviness" of the work, or the quantity and strength of the mordant on the cloth.

When removed from the first cistern, the pieces are well washed in water;* after which, they are winced in the weaker solution of substitute. This solution may be contained in a wince-pit or cistern capable of holding about three hundred gallons, with which quantity of hot water there should be mixed two quarts of substitute liquor and one quart of cleansing liquor. In this liquid twenty-eight or thirty pieces are winced for twenty or twenty-five minutes, at a temperature about 10° lower than the solution first applied. The second cistern requires to be renewed by the addition of two pints of substitute liquor and one pint of cleansing liquor for every twenty-eight pieces. Both this and the first cistern should be fresh charged every morning and emptied at night.

* If the work is heavy, it is also recommended to pass them between the squeezing rollers and again wash them in water.

The only remaining operation to which the pieces are subjected, previous to being dyed, is a thorough washing in water; and if the work is heavy, they should also be passed between the squeezing rollers, and again washed.

Where the use of dung is only partially superseded by that of the substitute, the pieces are sometimes first passed through the common dung emulsion and afterwards winced in a weak solution of substitute mixed with cleansing liquor or glue: or the pieces may be first passed in the ordinary manner through a mixture of half the usual quantity of dung with half the above proportions of substitute liquor and cleansing liquor, and be afterwards winced in a solution of substitute of the same strength as the second applied as above without any dung. For madder reds, the mixture of dung and substitute seems to be more advantageous than substitute or dung alone, but for madder purples and black a preference is given to the use of the substitute only.

The exposure to dung or substitute of cloths mordanted with alumina, should not be prolonged a sufficient time to allow of the union of the alumina with a full proportion of phosphoric acid; for colouring matters do not readily displace phosphoric acid from such a combination. The phosphate of iron, on the contrary, is easily decomposed by colouring matters.

In a few dyeing processes where it is of importance to avoid the aggregation of the particles of the mordant as much as possible, the pieces, instead of being exposed to dung or substitute, may be winced in a mixture of chalk and size with hot water. In this case, the chalk serves to fix the mordant on the cloth by withdrawing the small quantity of acid remaining in the subsalt; and the loosely combined mordant separated by the water is precipitated by the chalk, and thus rendered incapable of attaching itself to the fibre. If the goods contain an aluminous mordant, the process of wincing in chalky water should not be prolonged, and only a small proportion of chalk should be employed, as the precipitated alumina itself is apt to be removed by the action of an excess of chalk.

The dunging process is sometimes superseded by the operation of *branning*, which consists in wincing the goods in a mixture of bran and hot water. The action of bran is probably quite analogous to that of dung, the unprecipitated mordant dissolved by the water being separated from the liquid by the insoluble ligneous matter, while the undissolved mordant becomes more strongly attached to the cloth by combining with the mucilaginous and glutinous matters present, and also with the phosphoric acid of the phosphate of lime in the bran. The only cases in which branning is preferred to dunging are those in which the cloth is afterwards dyed to delicate shades of colour by means of cochineal and fugitive colouring matters.

After having been thus exposed to the action of either dung, substitute, chalk, or bran, the mordanted goods are ready to be exposed to the infusion of the dye-stuff; and in general, the sooner this is done, the better is the colour they assume.

The different vegetable colouring matters vary so

considerably in properties, that few observations of general application can be offered on the modes of preparing the various dyeing liquids. If the substance is very soluble, its solution may be made in the cold; but if only slightly soluble, heat may be applied, provided the colour is not deteriorated by exposure to a moderate heat. When a highly charged solution is required, (such as the topical and steam colours used in calico-printing,) concentration by evaporation is had recourse to; many vegetable colours, however, will not support a continued ebullition without losing something of their colour. If the goods are not kept in constant motion when in the dye-beck, the infusion should be freed from the insoluble ligneous matters by decantation or filtration; in some cases this operation may be avoided by enclosing the tinctorial matters in bags, which are withdrawn from the liquid when sufficient colour is imparted. But if the goods are kept in continual motion in the vegetable infusion, as is almost always done with cottons, the separation of the insoluble matters is unnecessary. The vegetable material is commonly introduced in a state of coarse powder into the dye-beck containing cold water; the pieces of mordanted cotton to be dyed are put in at the same time, and the temperature of the liquor is gradually increased by the introduction of steam.

In the dyeing of cottons, motion may be communicated to the goods, while in the dye-beck, by a wince or reel placed horizontally over the middle of the dyeing vessel, so that the cloth may be made to descend into either compartment of the dye-beck by the rotation of the reel. The dyeing vessel, which is commonly constructed of wood,* is represented in cross and longitudinal section at figs. 31 and 32:

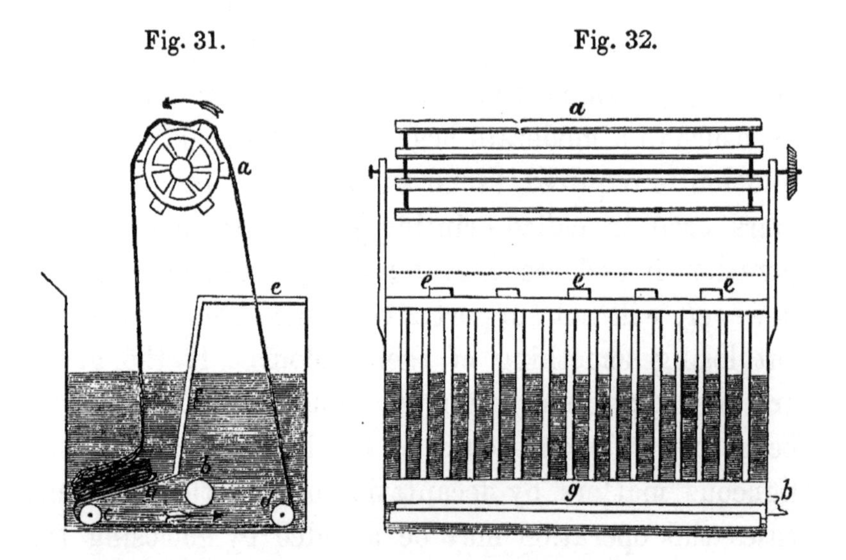

Fig. 31. Fig. 32.

a is the reel, containing six long wooden spars on its circumference; it is set in motion by being connected with one of the driving shafts of the factory. Steam is admitted to the vessel by the pipe *b*, the upper surface of which has a great number of small perforations. Twelve, eighteen, or twenty-four pieces of cotton, which are stitched together at their ends so as to form one endless web, pass over the reel in the direction of the arrows, and fall on a sloping iron ledge *g*, on one side of the vessel, from which they pass under the two rollers *c* and *d*. Four, five, or more of such endless webs may be set in motion by the same reel, they being prevented from entangling by wooden bars represented at *e*, which

* For madder work, dye-becks made of iron have been lately substituted for those of wood. When the metal is completely covered with oxide, it exerts no injurious action on the colouring matter.

reach from a cross-bar at the top of the vessel to the back. An inclined partition, *f*, extends through the whole length of the dyeing vessel, formed of several wooden spars placed a few inches apart from each other. The ordinary dimensions of the dyeing vessel are, six feet in length, four feet in width, and four feet in depth.

Such is a general view of the course of operations practised in the dyeing of goods with colouring matters which are naturally soluble in water, by the intervention of a mordant. If the colouring principle is insoluble in water, the mordant may be dispensed with; but it then becomes necessary to devise some means of obtaining such a solution of the colouring substance as will allow the deposition of the latter in its insoluble state, when a cloth impregnated with the solution is exposed to some chemical agent. This forms the third style of dyeing in the classification proposed at page 262. The principal insoluble vegetable colouring materials are indigo, safflower, and annatto, the nature of the processes for applying which to cloth has already been explained (page 243).

The only other style of dyeing which remains to be noticed is entirely different from either of the preceding; it is practised only on goods formed of the animal tissue, and admits of no more than one example in illustration. By this style, an orange colour is imparted to silk and wool, not from the solution of a colouring matter, but by effecting a certain chemical change in the fibre, through the action of dilute nitric acid. The orange colour is due to a substance formed by the decomposition of a portion of the silk or wool itself by the acid.

The temperature of the dye-beck at the time of dyeing depends almost entirely on the nature of the colouring matter. If it is readily attachable to the tissue, as indigo and colouring principles derived from metallic substances, for instance, and if easily altered by heat, as safflower, the dyeing solution may be used cold. But a hot liquid generally affords the most uniform colour, partly on account of the more ready disengagement of air-bubbles from between the fibres of the cloth. Dyeing with vegetable and animal colouring matters which require a mordant is also effected more rapidly with the assistance of heat, owing to the increased disposition of the mordant to unite with the colouring principle. In a few dyeing processes, however, where the mordant exists on the cloth in a soluble state when about to be dyed, a high temperature in the dye-beck is injurious, from the separation of a portion of the mordant from the cloth by the solvent action of the dyeing liquor. Hence it is that cotton, silk, and flax, impregnated with alum, absorb more colouring matter from some solutions at the ordinary temperature than at the boiling point. Where the operations are conducted on anything approaching a considerable scale, the most convenient and the most economical source of heat for the dye-beck is steam, which may be applied in three ways: 1°, by introducing it directly into the liquid by a pipe leading from the boiler; 2°, by causing it to circulate through a spiral pipe placed in the dye-beck; and 3°, by introducing it between the dye-copper and an exterior wooden case.

The vessels in which the dyeing decoctions are made and concentrated by evaporation, are usually of copper; for some delicate dyes, when a steam heat is applied, they are made of tin or of copper tinned inside. Copper boilers sometimes exercise considerable influence on the tints of the decoctions prepared in them, owing to the solution of some oxide of copper from the surface of the metal, by an acid existing either in the mordant or the dye-stuff. A solution of alum which has been boiled for some time in a copper vessel affords, with ammonia, a blue instead of a white precipitate; and wool acquires a greenish-grey tint when kept for some hours in a boiling solution of alum with cream of tartar contained in a copper vessel, which would not happen with the same solution in a vessel of tin.

In general, the vegetable and animal fibres become coloured much more readily when unspun than when wove into cloth. Wool in flocks, after having been washed, digested in an alkaline ley, and bleached by sulphurous acid, takes more colour than when spun into yarn, and the yarn more than when wove into cloth. This doubtless arises from the comparative difficulty with which the solution of the colouring matter obtains access to the internal fibres of the spun or woven tissue. The colour of the interior of a piece of thick woollen cloth dyed in the piece, is often less intense to the eye than the colour of the exterior. Certain disadvantages, however, sometimes attend the dyeing of wool in flocks and in thread: some colours, for instance, are susceptible of alteration in the subsequent manipulations in weaving; the texture of the fibre is sometimes altered so as to present inconveniences in these operations, and it is more expensive from the subsequent waste of some of the material.

The routine of finishing operations practised on cloths after being dyed, is varied considerably, according to the style of work and the nature of the stuff operated on. When the goods have remained a sufficient length of time in the dye-beck, they are removed and carefully washed in water to separate the coloured liquid retained mechanically between the fibres. The drying of the washed goods, if of silk and wool, is usually effected by exposure to the air at common temperatures; but occasionally heat is applied, the goods being introduced into a well ventilated apartment heated by the circulation of steam-pipes. The drying of goods dyed with delicate colours should always be performed in the shade.

The following account of the course of finishing operations practised on calico printed and dyed according to the madder style, will afford a general view of the treatment of cotton goods after having been dyed by means of a vegetable infusion with the intervention of a mordant. Some of the operations here noticed are unnecessary, however, in other styles of dyeing and printing.

Immediately contiguous to the dye-beck are usually placed two stone cisterns containing cold water, each surmounted by a reel, similar to that shewn in figs. 31 and 32. In one of these cisterns the cloth is washed as soon as it is taken out of the dyeing liquor, motion being communicated to the cloth by means of the reel. From the first cistern the pieces are transferred to the second, containing clean cold water, and from thence to a washing vessel of particular construction, called the *dash-wheel* (fig. 33).

Fig. 33.

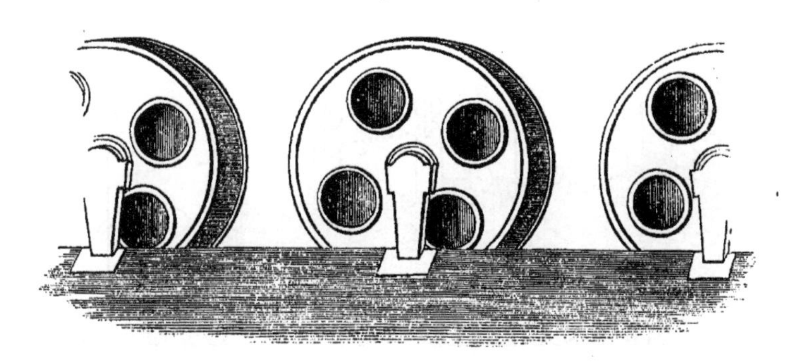

This is a hollow, circular, perpendicular wheel of five or six feet in diameter, and nearly two feet in depth, divided into four equal compartments by partitions proceeding from the axis to the circumference, each of which has a circular opening on one face of the wheel. Water is admitted into the compartments by a pipe concentric with the axis on which the wheel rotates. The pieces of cloth to be washed are put into the compartments through the circular openings in front, and water being admitted, the wheel is made to rotate rapidly, and thus wash the cloth with considerable agitation.

In the washing of cloths which require delicate treatment, as those dyed with fancy or spirit colours, for which the action of the dash-wheel is much too energetic, another washing apparatus is employed, called the *rinsing machine*, an idea of the ordinary construction of which will be afforded by the representation of its longitudinal section at fig. 34. It consists of a rectangular wooden cistern of from twenty to thirty feet long, three feet wide, and four feet high at one end, and three feet high at the other. The cistern is divided transversely into from six to

Fig. 34.

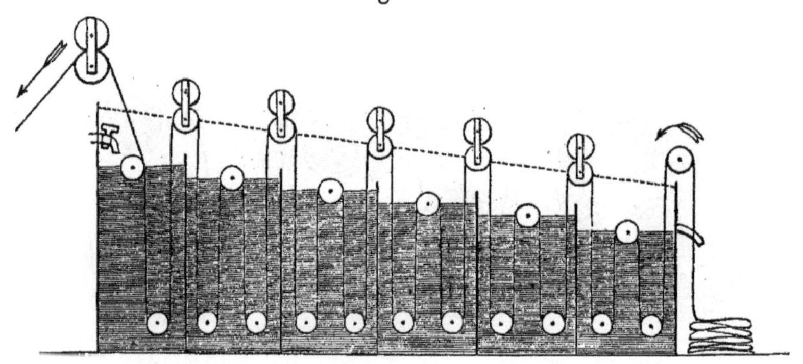

ten compartments, by partitions which gradually decrease in height from the higher to the lower end of the vessel. In each compartment except the highest, are placed three rollers, to regulate the passage of the cloth through the cistern, two of the rollers being near the bottom and the other at the top. Above each partition are placed two more rollers nearly in contact; and those above the higher end of the cistern and the first partition are squeezing rollers subject to considerable pressure, and worked by machinery connected with one of the driving shafts of the factory. The pieces of cloth to be washed are introduced into the cistern at the lower end, and traverse each compartment successively, being drawn through by the traction of the squeezing rollers at the upper end. A stream of clear water is made to flow into the cistern at the higher end and out at the lower, while the cloth is passing in the opposite direction; by which arrangement the cloth is brought successively into contact with purer portions of water and is discharged at the top perfectly clean. In the machine represented in the above figure, the water flows from one compartment into another through apertures near the tops of the partitions, and not over the partitions. In another form of the rinsing machine, the water passes from one compartment into the next through apertures at the tops and bottoms of the partitions alternately. It is to be observed that this machine is used only for goods which require more delicate treatment than is compatible with the dash-wheel or the wince-pit.

While the cloth is in the dye-beck, a considerable quantity of colouring matter attaches itself to the surface of the cloth, not in chemical combination with the mordant, but too strongly attached to be easily removed by washing in clean water. To get rid of this superfluous colour, the cloth, after having been washed at the dash-wheel, is winced either in a mixture of bran* and boiling water, containing about

* In effecting the removal of this excess of colouring matter, the most active constituent of the bran seems to be the husky part. The feebly combined colouring principle dissolved by the hot water and the mucilaginous matters present, instead of being retained in solution, is precipitated on the husky surface, and thus prevented from again attaching itself to the cloth. Coarse bran is better adapted for this purpose than fine, and flour seems to be altogether useless. An interesting memoir by M. Kœchlin-Schouch, on the use of bran in this operation (termed "clearing"), is contained in the ninth volume of the *Bulletin de la Société Industrielle de Mulhausen.*

a bushel and a half of wheat bran for every ten pieces of calico, or else in a dilute solution of soap. The addition of a little caustic alkali to the soap or bran is sometimes made; but neither an alkali nor soap can be used for this purpose without great care, as the tints of all vegetable colouring principles are slightly deteriorated by these agents. For most vegetable colouring matters besides madder, bran only is admissible; and even in bran-water, the wincing sometimes must not exceed a few minutes. With madder colours only, the wincing may be continued for from ten to twenty minutes.

The complete removal of the superfluous colour from a piece of cloth which is to present a white pattern is generally effected, when madder is the only vegetable colouring matter present, by wincing the cloth for a few minutes in a solution of chloride of lime, not stronger than 3° Twaddell (1015). This operation usually follows that of branning or soaping, but sometimes the branning is altogether omitted when the solution of chloride of lime is employed.

Few vegetable colouring matters, however, can be exposed to the action of chloride of lime without considerable deterioration; hence, when other dye-stuffs than madder are employed, the "clearing" of the dyed cloth is effected, sometimes by exposure to air and light, but the process of branning or soaping is generally found to be sufficient of itself.

After having been thus cleared of the redundant colour, the cloth is washed, and then submitted to an operation for expelling almost the whole of the water it contains; which consists either in passing it between two rollers revolving against each other under considerable pressure (squeezing rollers), or else in rotating the cloth so rapidly as to cause the water to be driven out by the centrifugal force thus excited. One of the machines used for the latter purpose is represented in perpendicular section across the centre at fig. 35, and as viewed from above in fig. 36: *a* and *b* are two copper cylinders connected together at bottom so as to form one vessel, which rotates with the axis *c*. These cylinders are enclosed in a wooden case *d*, which is in communication,

Fig. 35.

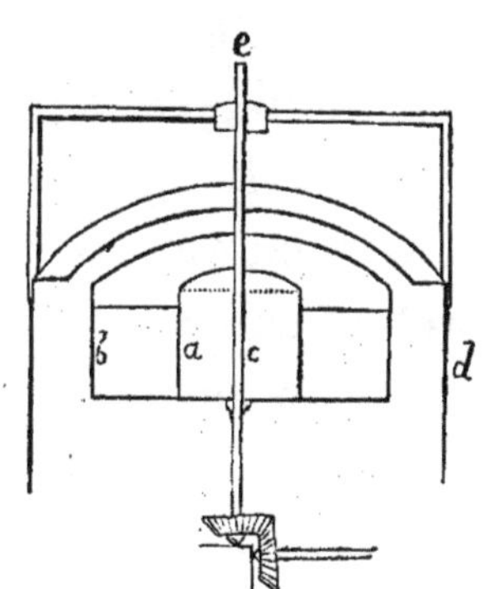

Fig. 36.

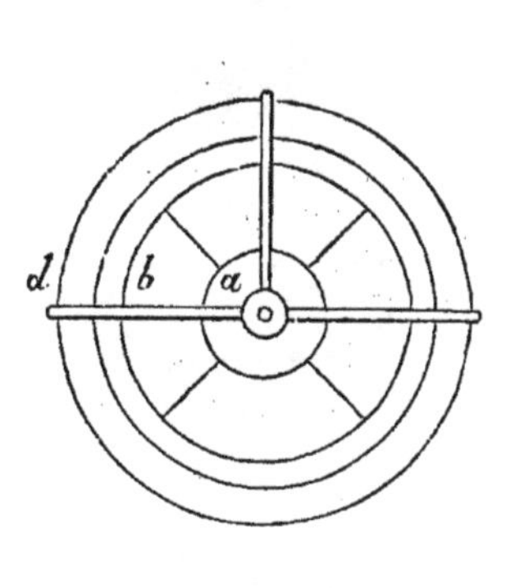

at bottom, with a drain or gutter. The cylinder *b* has a great number of small perforations, and is divided by partitions into four equal compartments. The wet cloth which is to be dried is placed in the compartments between the two cylinders, and the apparatus is rotated with a velocity of nine hundred or a thousand revolutions per minute; the water is thereby driven from the cloth through the perforations in the cylinder *b* to the outer case, from whence it flows out by a gutter or drain. After a few minutes the cloth becomes nearly dry, and when the machine is opened, is found to be strongly compressed against the perforated cylinder.* In another form of this machine, which works with much less noise than the preceding, the cylinders are arranged vertically, so as to form an apparatus somewhat resembling the dash-wheel (fig. 33, page 319).

When the cloth has been thus far dried, either by the squeezers or the "water extracter" just described, it is folded evenly and then passed, in a length of ten pieces, through a mixture of blue starch and water.

Fig. 37.

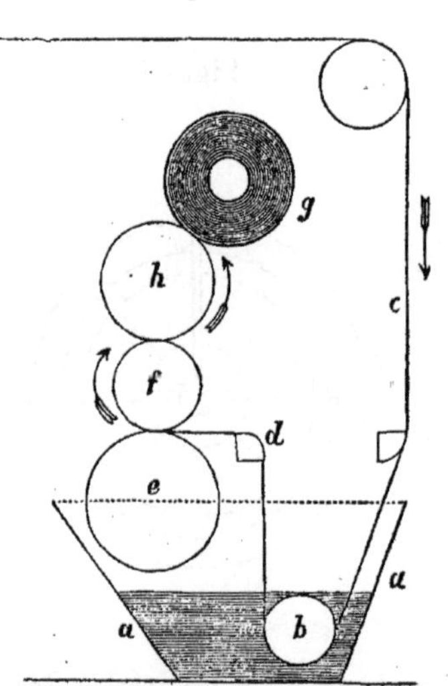

A cross section of the *starching machine* is represented in fig. 37: *a* is a wooden trough to contain the mucilaginous liquid; *b* is a small cylinder revolving in the liquid; around this is passed the web of calico *c*, which is then drawn over a fixed brass bar *d*, with diagonal notches on its front, for the purpose of removing creases from the cloth; *e* is a wooden cylinder covered with cloth, revolving in close contact with the brass cylinder *f;* the calico is passed between these cylinders to be freed from the superfluous starch, and is then rolled off upon the cylinder *g*, the axis of which is not fixed, but so contrived as to recede gradually from the wooden cylinder as the roll of calico increases in diameter.

After being starched, the ten pieces of calico are

* This machine is known by the name of "water extracter."

passed through the *steam drying machine*, which consists of several hollow copper cylinders, each about twenty inches in diameter and three feet in length, fitted up with machinery by which all the cylinders or drums may be rotated together at the same velocity. Steam is admitted to the drums through stuffing boxes at one end of the axes, and at the other end are placed pipes to discharge the condensed water. The number of drums arranged together in one system varies from five to thirteen, according to the quantity of work required; they are sometimes placed in one line, but usually in two lines, one immediately over the other (as fig. 38), with the cir-

Fig. 38.

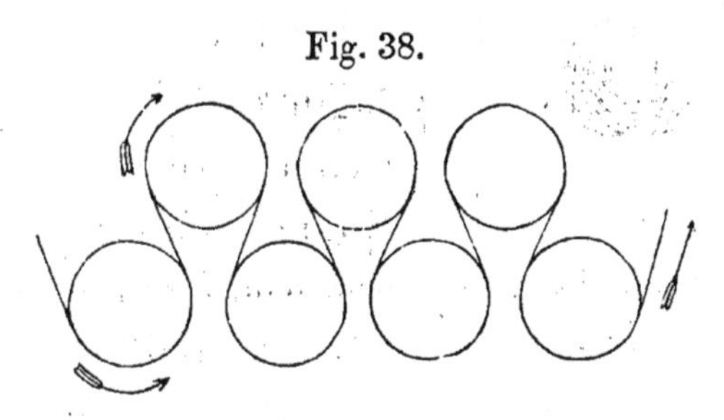

cumference of one drum distant about four or six inches from that of the next in the line of the axes of the drums. The calico passes through the machine in the direction of the arrows in the above figure. The drums are surmounted by a hood and flue for the purpose of conducting the steam out of the chamber.

The last finishing operation to which dyed and printed cottons are sometimes subjected is *calendering* or glazing, the object of which is to make the surface of the cloth smooth, compact, and uniform. This is effected by passing the piece between two cylinders revolving in such close contact that their pressure gives the cloth the appearance of having been ironed. One of the cylinders is made of iron, and is hollow for the purpose of admitting steam or a hot iron rod, when the application of heat is necessary. The material of the other cylinder was formerly wood, but for some years past pasteboard has been very generally substituted. The cylinder of paper has several decided advantages over that of wood. It takes a finer polish, it has no tendency to crack or warp, and from having a certain degree of elasticity it gives a more equal pressure on all parts of the cloth than could be applied by a wooden cylinder. The paper cylinder is constructed by placing circular discs of stout pasteboard on a square bar of wrought iron as an axis. The external discs are of cast iron, a little less in diameter than the remainder of the cylinder. The discs being screwed down tight, the cylinder is placed in a stove, and kept for several days at as high a temperature as the paper will bear without being charred or rendered very brittle. As the moisture is driven off, the pasteboard shrinks, and the screws must be tightened to keep the mass as compressed as possible. When no further diminution in bulk is perceived, the cylinder is removed from the stove, and carefully turned on a lathe. The diameter of the paper cylinder is usually fourteen inches, and that of the opposed iron cylinder eight inches. Four or five cylinders are commonly arranged together one over the other on the same frame.

The glazing of calicos was formerly executed by the hand with a hot iron, at an expense of about one shilling per piece of twenty-eight yards; the cost of glazing by machinery as above is from threepence to sixpence per piece.

The purity of the water employed in dyeing operations is a subject which deserves the especial attention of the practical dyer. The finest colours are in almost all cases obtained by making use of distilled water, that being free from all earthy impurities. Rain water and the water of an Artesian well are, in general, better adapted for dyeing than spring water and river water, as the latter contain in solution a quantity of lime, which sometimes falls down in combination with the colouring matter as an insoluble precipitate, occasioning a considerable loss of dye-stuff. Spring and river water also generally contain a sensible quantity of iron, which always communicates a brown tinge to the goods washed in such waters.

When the yellowish Dutch madder is boiled with pure distilled or rain water, the residuary ligneous matter has a light brown colour, and imparts only a faint red colour to a boiling solution of alum. When, on the other hand, spring water is substituted, the residue is dark reddish-brown, and a solution of alum in which it is boiled becomes of a dark red colour. In the first case, the quantity of madder red remaining in the residue is much less than in the second. The madder red at first dissolved is precipitated by the lime of the spring water, imparting to the residue its dark colour, and is dissolved by the boiling solution of alum. Hence pure water dissolves more madder red than water holding lime in solution. Similar results are obtained with Fernambouc wood and logwood. (Dr. F. Runge, *Farben-chemie.*)

In some print-works in Lancashire distinguished for their fancy styles, it is a common practice to add a little dilute sulphuric acid to the water, if the latter contains carbonate of lime. The sulphuric acid converts the carbonate into sulphate of lime, which scarcely affects the brilliancy of the colours of the dyed or printed goods. It is of importance that there should be no excess of the acid. When cochineal colours are washed, distilled water is usually employed; but where this cannot be readily obtained in sufficient quantity, water treated with acid, as above, is used. These remarks are applicable to water containing calcareous matters only.

Dr. Clarke's process for purifying water from carbonate of lime has not yet been introduced into the Lancashire print-works, but if efficiently conducted, it would no doubt be found highly advantageous.

Water which infiltrates marshy ground often contains in solution a quantity of decomposed vegetable matter, which is also very detrimental to certain colours. Not only is the shade of colour modified by the attachment of the organic matter, but certain metallic colouring materials, especially chrome-yellow and chrome-orange, are decomposed and converted into a brownish-black substance through the action of the organic matter. This proceeds from the generation of soluble earthy or alkaline sulphurets through the decomposition of the soluble sulphates which spring water always contains; the blackening of the chrome-yellow and chrome-orange is due to the formation of sulphuret of lead by the action of the soluble sulphuret thus produced.

A simple and efficacious method of rendering hard water well adapted for dyeing operations is practised at the Dukinfield branch of the Mayfield print-works, Manchester, on all the water consumed there, which amounts to six or eight hundred thousand gallons daily. It merely consists in mixing the refuse of the madder dye-becks with the water; the remaining colouring matters of the madder then precipitate the iron and lime in an insoluble form, and the water is obtained clear and fit for use by allowing the precipitate to settle in a large reservoir, and then filtering the water through a bed of gravel.

At an extensive silk-dyeing establishment in London, the only water employed is that raised from an Artesian well.

In one dyeing process, however, namely, the production of a black colour by means of infusion of galls, valonia, or sumach, and copperas, the water which is preferred by some dyers is hard spring water. To produce in a liquid a given depth of colour, distilled water requires more dye-stuff than common spring water. This is illustrated in the following experiment devised by Mr. Phillipps. Into two glass jars of the same size, each half filled with distilled water, introduce equal quantities of infusion or tincture of galls or sumach, and an equal number of drops (only three or four) of a solution of copperas. A faint purplish colour will be developed in both jars; but if one is filled with spring water, the colour in that rapidly becomes dark reddish-black, and one-half more water is required to reduce it to the same shade of colour as the other. The water which is found by experience to be best adapted for dyeing with galls and sulphate of iron differs from distilled water in containing sulphate of lime, carbonate of lime held in solution by free carbonic acid, and chloride of calcium. The beneficial ingredient seems to be the carbonate of lime, which possesses slight alkaline properties; for, if the smallest quantity of ammonia, or of bicarbonate of potash, is added to the distilled water in the above experiments, the purple colour is struck as rapidly and as deeply as in the spring water; chloride of calcium and sulphate of lime, on the contrary, produce no sensible change either in the depth of colour or the tint. The effect is no doubt referable to the action of the alkali or lime on the protosulphate of iron, by which the sulphuric acid of the latter is withdrawn, and hydrated protoxide of iron set free; for protoxide of iron is much more easily peroxidized and acted on by tannic and gallic acids (the dyeing principles of galls) when in the free and hydrated state, than when in combination with sulphuric acid. Neither the caustic fixed alkalies (potash and soda) nor their carbonates can be well introduced in the above experiments, as the slightest excess reacts on the purple colour, converting it into a reddish-brown. Ammonia, lime-water, and the alkaline bicarbonates also produce a reddening, and if applied in considerable quantity, a brownish tinge.

But the dyeing operations in which hard water

is preferable to soft are so few in number, that the generality of the above statement concerning the superiority of soft water is scarcely at all affected.

§ IV. CALICO-PRINTING PROCESSES.

Although the different methods of procedure in the printing of cottons are almost as numerous as the different kinds of patterns which may be produced, yet each colour in a pattern is always applied by one of six different styles of work, by the proper combination of two or more of which the cloth may be ornamented with any pattern, however complicated. These styles are quite distinct from one another; each requires a peculiar process and a different manipulation.

The six styles alluded to are the following:

1. *Madder style, for soluble vegetable and animal colouring materials.*—In this kind of work, which derives its name from being chiefly practised with madder, the thickened mordant is first imprinted on the white cloth in patterns, and after the cloth has been aged and dunged, the colour is imparted by passing the cloth through the dye-beck. On those portions of the cloth on which the mordant is applied, the colouring matter attaches itself in a durable manner, but on the unmordanted portions the colour is feebly attached, so that it may be wholly removed by washing either in soap and water, in a mixture of bran and water, or in a dilute solution of chloride of lime.

2. *Topical style, for steam and topical colours.*

—Such colouring matters as are incompletely, or not at all, precipitated from their solutions on being mixed with certain solutions of a mordant, are sometimes printed on the cloth with the mordant, and the fixation of the colour is afterwards effected by exposing the cloth to steam. Some colouring matters applied topically in a state of solution become firmly attached to the cloth without a mordant and without the process of steaming, but merely by drying with exposure to the air.

3. *For mineral colours* (*padding style*).—To produce a figure in a mineral colouring material the cloth may be first printed with one of the two saline solutions, and be afterwards uniformly impregnated with the other. To obtain a ground of a mineral colour, one or both of the solutions may be applied by the padding machine.

4. *Resist style.*—In the processes referable to the resist style, the white cloth is first imprinted with a substance called the *resist*, or *resist paste*, which has the property of preventing those portions of the cloth on which it is applied from acquiring colour when afterwards exposed to a dyeing liquid. Resists are divisible into two classes; one is employed to prevent the attachment of a mordant, and the other that of a colouring matter.

5. *Discharge style.*—The object of the processes belonging to this style of work is the production of a white or coloured figure on a coloured ground. This is effected by applying topically to the cloth already dyed or mordanted, a substance called the *discharger*, which has the property of decomposing or dissolving out either the colouring matter or the

mordant. Chlorine and chromic acid are the common discharging agents for decomposing a vegetable or animal colouring matter, and an acid solution for dissolving a mordant.

6. *For China blue.* — This is a very peculiar style, and is practised with one colouring matter only, namely, indigo. This pigment is printed on the cloth in its insoluble state, and is dissolved and transferred to the interior of the fibre by the successive application of lime and copperas, with exposure to the air.

The topical application of the colouring matter, mordant, discharge, or resist, may be made by five different methods:

1. The simplest is by means of a wooden block, of from nine to twelve inches in length, and from four to seven inches in breadth, bearing the design in relief as an ordinary woodcut; or, when the design is complicated, and a very distinct impression is required, the figure is sometimes formed by the insertion of narrow slips of flattened copper wire, the interstices being filled with felt. The block is worked by the hand, and is made of sycamore, holly, or pear-tree wood, on a substratum of some commoner kind of wood. It is charged with colour or mordant by pressing it gently on a piece of superfine woollen cloth, called the sieve, which is kept uniformly covered with the thickened colouring matter or mordant by an attendant boy or girl, called the "tearer," (corrupted from the French word *tireur*,) who takes the colour up by a brush from a small pot and applies it evenly to the woollen cloth. This cloth is stretched

tight over a wooden drum, which floats in a tub full of old paste or thick mucilage to give it sufficient elasticity to allow every part of the raised device on the block to acquire a coating of colour. The calico being laid flat on a table covered with a blanket, the charged block is applied to its surface (the printer being guided where to apply the block by small pins at the corners) and struck gently to transfer the impression. The application of the block to the woollen cloth and the calico alternately is continued until the whole piece of calico is printed. By the ordinary method, a single block prints only a single colour; hence, if the design contain five or more colours, and all be printed by block, five or more blocks will be required, all equal in size with the raised parts in each corresponding with the depressed parts in all the others.

If the design, however, requires different colours to be applied in figures in straight and parallel stripes, all the stripes may be applied by one block at a single impression, and the block is also charged with the different colours by a single application to the surface of woollen cloth. The colours to be applied are contained in as many small tin troughs as there are colours, arranged in a line. A little of each colour is transferred from the troughs to the woollen cloth by a kind of wire brush consisting of wires fixed in a narrow piece of wood. The colour is distributed evenly in stripes over the surface of the sieve, by a wooden roller or rubber covered with fine woollen cloth. For the rainbow style, the colours are blended into one another at their edges by a brush or rubber drawn to and fro in a straight line.

An important improvement in the construction of the hand-block has been recently adopted in most well conducted print-works, which consists in the application of a stereotype plate as the printing surface. To make the stereotype plate, a model is first formed from the pattern, about five or six inches in length, and from an inch and a half to five inches in width, according to the design. A mould is produced by stamping from the model; and from the mould, fixed in a block, the stereotype copies are produced in a mixed metal, composed of eight parts of bismuth, five parts of lead, and three parts of tin. When a sufficient number of the pieces is prepared, their surfaces are filed down, and they are then fixed to a stout piece of wood.

2. The hand-block has been superseded to a great extent on most parts of the Continent by a machine called the Perrotine, in honour of M. Perrot, of Rouen, its inventor. This machine executes block-printing by mechanical power, and is intermediate in its mode of working between block-printing and cylinder-printing, to be noticed immediately. The perrotine is composed of three or four wooden blocks, from two to five inches broad, and as long as the breadth of the cloth to be printed. The blocks are faced with pear-tree wood and engraved in relief. They are mounted in a cast-iron frame with their planes at right angles to each other, and by a simple contrivance are charged with a coat of coloured paste and then pressed successively against the cloth to be printed. The cloth is drawn by a winding

cylinder between the engraved blocks and a square prism of iron, mounted so as to revolve on an axis against the blocks. Two or three only of these machines are in operation in this country.

3. About the commencement of the present century the hand-block and flat copper-plate, till then the only means of impression possessed by the printer, began to be superseded, for most styles of work, by cylinders of engraved copper. A general idea of the nature of this mode of printing may be conceived with the assistance of the annexed figure; *a* repre-

Fig. 39.

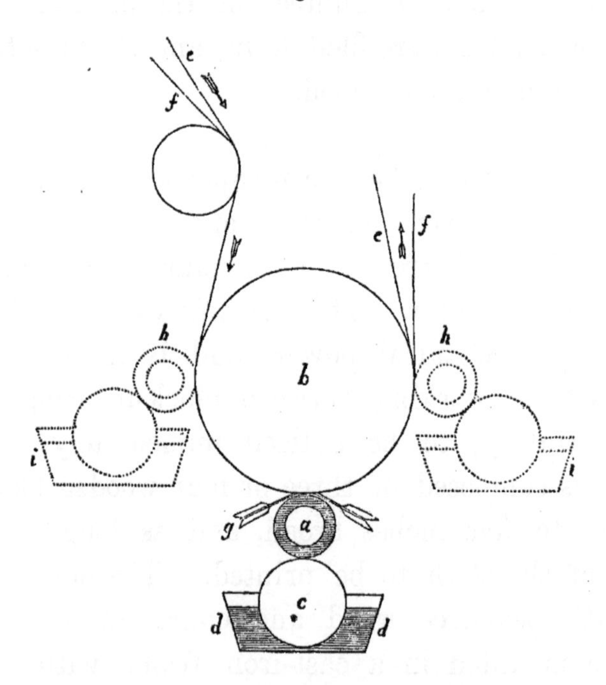

sents the engraved cylinder or roller, mounted on a strong frame-work, so as to revolve against two other cylinders *b* and *c*. The cylinder *c*, which is covered with a woollen cloth, dips into the trough *d*, containing the solution of the colouring matter or mordant properly thickened, and thus acquiring itself a coating of the colour imparts some of it in the act of rotation to the engraved roller *a: b* is a large iron drum or cylinder, the surface of which is rendered elastic by several folds of woollen cloth ; around this drum travels an endless web of blanket-stuff, *e*, in the direction of the arrows, accompanied by the calico passing between it and the engraved cylinder. The pressure of the cylinders against each other is regulated by screws or levers, which can be tightened or slackened at pleasure.

The excess of colouring matter or mordant which is communicated to the engraved roller by the cylinder *c* must obviously be removed before it comes into contact with the calico; this is accomplished by scraping the surface of the roller as it revolves, by a sharp-edged plate, usually of steel, called the *colour doctor* (*g*). Another similar plate is placed on the opposite side, called the *lint doctor*, the office of which is to remove the fibres which the roller acquires from the calico. With some colour mixtures and mordants, those containing salts of copper for instance, doctors composed of gun metal, bronze, brass, and similar alloys, are substituted for those of steel, as the latter would become corroded through the chemical action of the mordant or colour mixture.

Such is the method of printing calicos by the roller for a single colour; but the mordants or mixtures for two, three, or even eight colours may be applied at the same time by having as many engraved rollers with their appendages revolving simultaneously against the iron drum, as represented in fig. 39, by the dotted cylinders and troughs *h*, *h*, *i*, *i*. Extreme nicety of arrangement is required to bring all the rollers to print the cloth at the proper places, but when once properly adjusted each may be made to deposit its colour or mordant on the calico with the greatest certainty and regularity.

The diameter of the printing roller varies from four or six inches to a foot or even more; its length varies from thirty to forty inches, according to the breadth of the calico to be printed. It was formerly made of plates of copper hammered into a circular form and joined by brazing; but as the engraving easily gives way on the brazed joint, the roller is now bored and turned from a solid piece of metal. The engraving is not commonly etched by the ordinary graver, as was formerly done at a great expense, but by the pressure of a steel roller, called the *die*, from three to four inches in length (according to the pattern), containing the figures in relief which it imparts in intaglio to the softer copper. The steel die is made in a similar manner by powerful pressure against another steel roller called the *mill*, of similar size, which is engraved by the common graver while in the soft state, and afterwards hardened by being heated and then plunged into cold water. The steel die to receive the figure in relief is also in the soft state when pressed against the hardened engraved mill, and is itself hardened before being applied to the copper roller. The cost of engraving a roller in this manner is very little more than one-eighth that of engraving by the hand.

For some peculiar styles of pattern, the copper roller is etched instead of being engraved by indentation. The roller being heated by the transmission of steam through its axis, is covered with a thin coat of resist varnish, and when it is cold, the pattern is traced with a diamond point by a very complicated and ingenious system of machinery, the roller being slowly revolved at the same time in a horizontal line beneath the tracer. After having been etched on its whole surface, the roller is suspended for about five minutes in a trough containing dilute nitric acid, which dissolves the copper in the lines exposed by the removal of the varnish, but the parts still covered remain unacted on. The importance and value of this method arises from its affording an endless variety of curious configurations, which can hardly be copied or even imitated by the hand engraver.

The following ingenious method of imparting a printing surface to a copper roller has been extensively practised of late in one of the best conducted print-works in Lancashire. It is only applicable to rollers to be used for printing a full ground, sprigs or other designs being left blank, for grounding in other colours if required by the block, at a subsequent operation.

The copper roller is, in the first place, painted with a resist varnish on its whole surface, with the exception of the figures to be left blank; and to render the blank parts perfectly clean, the roller is dipped, first, into weak nitric acid, and immediately afterwards, into clean water. From the water, the roller is transferred to a solution of sulphate of copper and placed in connection with a galvanic battery, whereby it acquires a coating of copper on the designs where the varnish had not been applied. These raised designs are afterwards polished smooth, so that when the roller is in use, they become perfectly cleared by the "doctor," and the ground only is imprinted on the cloth. It will be observed that this method of obtaining a printing surface is essentially different, in principle, from the etching process just described; in one method, the surface of the roller is covered with varnish on the parts to be raised, and in the other, on the parts to be depressed.

4. A very ingenious method of printing has been lately introduced, distinguished as "press printing," by which block printing with several different colours may be executed at one impression. A sketch of the principal parts of the press-printing machine is shewn in fig. 40. The block itself *a* consists of a well prepared tablet of wood, about two feet six inches square, supported in an iron frame in such a manner that it can be raised or lowered vertically at pleasure. The face of the block is divided into as many stripes (crossways with the table) as there are colours to be printed, which we may suppose, for illustration, to be five. The stripes are about six inches in breadth and as long as the breadth of the cloth to be printed; each one prints a different colour, and the whole five form together the combined pattern. The printing surfaces are stereotype casts, made of the mixed metal, bismuth, tin, and lead (see page 335).

The mode of applying the colours to the printing surface is very ingenious. At the bottom of the wooden frame *b*, near to one end of the table, is a felt cushion about the same size as the entire block; and immediately within one side of the frame are

Fig. 40.

arranged in a line five little troughs (or as many as there are colours to be printed), containing the thickened colours. By means of a long piece of wood, so formed as to dip into all the troughs at once, the attendant "tearer" applies a little of each of the five colours to the surface of felt, over which the colours are evenly spread by a brush in five stripes without any intermixture. The breadth of the stripes is the same as the breadth of the stereotype rows on the block.

The cushion being thus charged, the frame is slid forward on the table on a kind of railway, until it lies immediately underneath the block, which is then lowered by the "pressman" upon the felt cushion, whereby each of the five stripes on the block becomes charged with its proper colour. This being done, the block is raised, the colour-frame withdrawn, and the block caused to descend on the cloth, which it imprints in five rows with different colours. When the block is raised, the cloth is drawn lengthways over the table about six inches, or exactly the width of one stripe on the block: the "tearer" again slides over the cushion with more colour, and the block is again charged and applied to the cloth. As a length of the cloth equal to the width of a stripe is drawn from underneath the block at each impression, every part of the cloth is brought into contact successively with all the stripes on the block. The part printed by the fifth stripe at the first impression becomes printed by the fourth stripe at the second impression, by the third stripe at the third impression, by the second stripe at the fourth impression, and by the first stripe at the fifth impression. When this machine is well managed, its action is very neat; but extreme nicety is required in properly adjusting all the moving parts of the press in order to prevent confusion of the colours and distortion of the pattern.

5. The only mode of printing which remains to be noticed is "surface printing," which is merely a modification of roller printing, the cylinder being made

of wood instead of copper. The pattern is either cut in relief, as in the ordinary block, or it is formed by the insertion, edgeways, of flattened pieces of copper wire. This cylinder is mounted in a frame as the copper roller, and is supplied with colour by revolving against the surface of an endless web of woollen cloth, which passes into a trough containing the colour or mordant. Surface printing is scarcely at all practised in this country, but in certain styles of work it presents some advantages over copper roller printing, particularly where substances which corrode copper, but not wood, are to be applied. It is practised more extensively in Ireland than in Lancashire.

Thickeners.—The thickening of the solution of the mordant or the colouring matter in order to prevent the liquid from extending beyond the proper limits of the design, is a subject which requires considerable attention in the successful practice of calico-printing. The degree of consistency and the nature of the thickening material require to be varied according to the minuteness of the design and the nature of the substance to be applied, for particular colouring matters and particular mordants often require particular thickeners. Two similar solutions of the same mordant, equally thickened, but with different materials, afford different shades of colour when dyed in the same infusion;* and the time required for the fixation of the mordant during the

* Solutions of salts of iron or copper thickened with starch give a deeper colour to the cloth, when afterwards dyed, than the same solutions if thickened with gum arabic.

ageing is considerably affected by the nature and consistence of the thickening material with which the mordant had been applied.

The following is a list of the thickening materials commonly employed:

1. Wheat starch.
2. Flour.
3. Gum arabic.
4. British gum.
5. Calcined potatoe starch.
6. Gum senegal.
7. Gum tragacanth.
8. Salep.
9. Pipe-clay, mixed with either gum arabic or gum senegal.
10. China clay, mixed with gum arabic or senegal.
11. Dextrin.
12. Potatoe starch.
13. Rice starch.
14. Sago, common and torrefied.
15. Sulphate of lead, mixed with gum arabic or senegal.

The most useful thickeners are wheat starch and flour. When either of these or any kind of starch (not roasted) is employed, the mixture with the mordant or colouring matter requires to be boiled over a brisk fire for a few minutes in order to form a mucilage; the consistency of the mixture, when cold, diminishes if the ebullition is continued for a longer time. Neither flour nor any kind of unroasted starch is well adapted for thickening solutions containing a free acid or an acidulous salt; if other circumstances, however, should render the introduction of another thickener inadmissible in such a case, the acid or acid salt is always mixed with the thickening after the latter has been boiled and cooled to 120° or 130° Fahr. If the acid is boiled with the mucilage, the mixture completely loses its consistency.

Starch is almost the only thickener employed for mordants containing no free acid, and the mordant seems to combine with the stuff more readily when thickened with starch than when thickened with gum.

During the ebullition of starch with red liquor, a precipitate of subsulphate of alumina is produced (see page 283); but this precipitate is completely redissolved as the mixture cools, its solution being apparently facilitated by the starch.

Next to wheat starch and flour, the most generally useful thickener is gum arabic. With this substance, however, many metallic solutions, such as those of salts of tin, iron, and lead, cannot be well employed, as such solutions cause the formation of precipitates with an aqueous solution of gum. This objection to the use of gum does not apply to so great an extent to salts of copper.

The lime which is contained in all gum arabic met with in commerce is apt to affect the light shades of some colouring matters; but this inconvenience may be overcome by adding to the gum a small quantity of oxalic acid, which converts the lime into the insoluble oxalate.

Gum senegal is used for the same purpose as gum arabic.

British gum, torrefied or calcined farina, dextrin, and torrefied sago starch (known as "new gum substitute"), are intermediate, both in their properties and applications, between common starch and gum arabic. Calcined potatoe starch is chiefly used with solutions applied by the padding machine, which require very little thickening.

Gum tragacanth and salep are commonly employed as thickeners for solutions of salts of tin and for mixtures containing a considerable quantity of a free acid. Salep does not stiffen and harden the stuffs so much as most other thickeners, and is hence found advantageous for mixing with topical colours. It gives considerable consistence to water, but the mixture is apt to become thin on standing. It is remarkable that a mixture of solutions of gum tragacanth and gum senegal, both of the same strength, possesses only one-half or one-third the consistency of the two solutions before being mixed.

Pipe-clay, China clay, or sulphate of lead, when mixed with either gum arabic or gum senegal, is also used with acid mixtures, and with solutions of salts of copper when applied as resists for the indigo vat. The earthy basis acts as a mechanical impediment to the attachment of a colouring matter, when the latter is applied to the whole surface of the cloth.

When the mordant to be printed is colourless, or nearly so, as alum, red liquor, and salts of tin, it is mixed with a little decoction of logwood, Brazil wood, or some other fugitive dye, in order to render the design on the cloth more perceptible. This addition of colour is called *sightening.*

We proceed, in the next place, to consider some particular examples of printing processes in illustration of the six different styles of work noticed at page 331.

I. MADDER STYLE.

The madder style is applicable, not only to the dye-stuff from which it derives its name, but to nearly all organic colouring materials which are soluble in water, and capable of forming insoluble compounds with mordants, and is much more extensively practised than any other style.

The ordinary course of operations to which a piece of cotton is subjected in order to be printed and dyed according to this style is the following:

1. Printing on the thickened mordant, which is commonly done by the cylinder machine.

2. Immediately after the imprinting of the mordant, the cloth is dried by being drawn either through the hot-flue,* or over a series of thin sheet-iron boxes, heated by means of steam, and is then conducted into the "ageing" room, where it is suspended, free from folds, for one or two days, according to the nature of the mordant and the temperature. The ageing room should not be very dry, or heated above the ordinary temperature, except during winter.

During this suspension, the greater part of the mordant undergoes a chemical alteration, by which it becomes attached to the cloth in an insoluble state.

* The hot-flue is a long gallery or passage, commonly heated by the flue of a furnace at one end, which runs through the whole length of the gallery on its floor. It is advantageous to have the upper surface of the flue formed of rough cast-iron plates, which become quickly heated and present a good radiating surface. A piece of calico (28 yards) is usually drawn through the flue in about two minutes.

Red liquor and acetate of iron part with a portion of their acetic acid; the former affords a deposit of subsulphate of alumina, the latter of subacetate of iron, as before explained; and the aluminate of potash affords a precipitate of alumina, through the action of the carbonic acid in the atmosphere (see page 285). Annexed is a specimen of calico in this

No. 1.

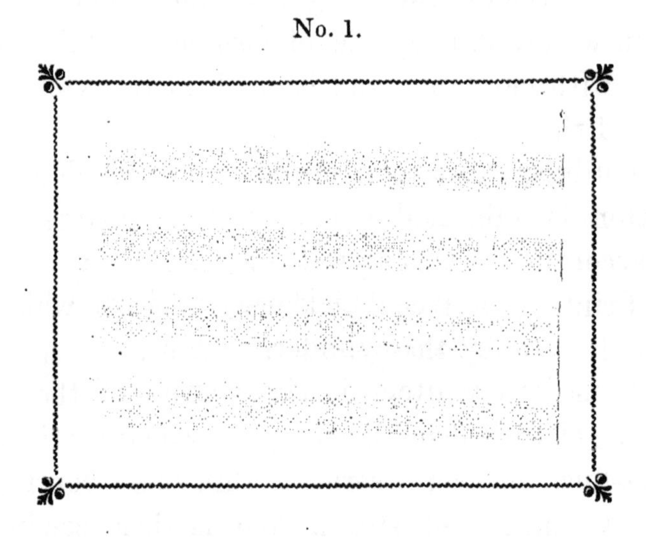

stage of the process, the mordant on which is red liquor, "sightened" with a little decoction of peach-wood or similar dye-stuff.

3. After having been suspended in the ageing room for a sufficient time, the printed cloth is drawn through the dung-becks, or else through a strong solution of dung substitute (page 307), whereby a part of the undecomposed mordant is separated from the cloth and prevented from acting on those parts which had not been printed, the thickening paste is removed, and the mordant remaining on the cloth becomes more strongly attached, by uniting with some of the constituents of the dung or of the substitute.

When taken out of the dung-becks, the cloth is immediately washed in a cistern of cold water, or sometimes both squeezed and washed, and then commonly winced for twenty or twenty-five minutes in a weak solution of substitute and size, by which the fixation of the mordant on the cloth is rendered complete. If the mordant is white, the cloth presents little trace of the design when taken from the substitute; the specimen No. 2. shews the appearance which

No. 2.

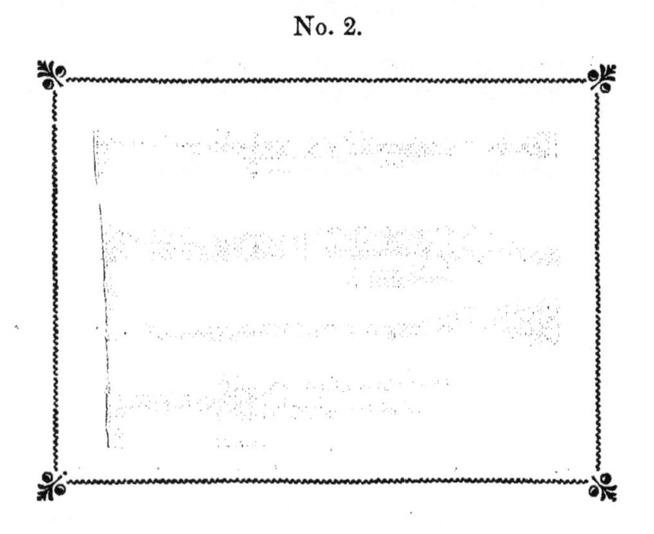

No. 1. acquires by being thus treated. (For an account of the manner of performing this operation, and of the probable action of the dung, see page 306.)

4. After having been well washed, the cloth is ready to be exposed to the dyeing liquor, in which it is kept for two or three hours, being constantly turned by a wince from one compartment of the dye-beck to the other. With madder and some other colours the goods are introduced into the cold mixture of water and ground dye-stuff, and heat is then gradually applied by the introduction of steam, until

the temperature of the liquid is very near ebullition (see page 313). When taken out of the dye-beck and simply washed in cold water, the cloth has the appearance of the specimen No. 3.

No. 3.

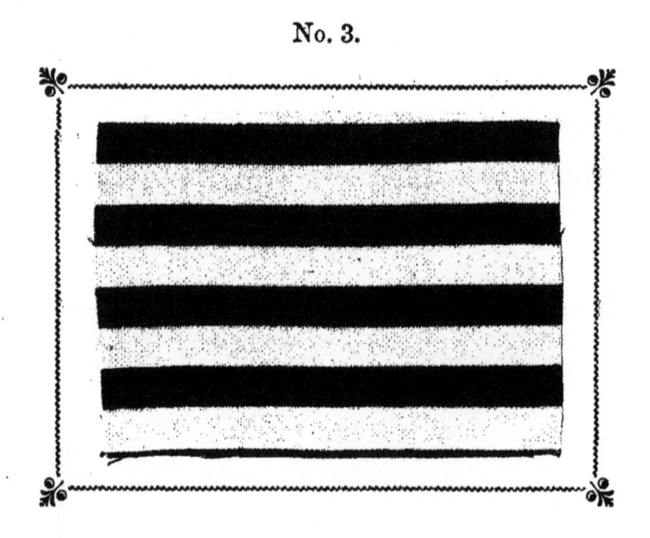

5. The next process to which the cloth is subjected is the "clearing," or the removal of the excess of colouring matter loosely attached to the exterior of the fibres. The processes for this purpose are varied with different dye-stuffs, and even with different varieties of the same kind of dye-stuff, according to their fixity (see page 321). With madder colours, where no sumach is employed, the goods may be cleared in the following manner, having been previously well washed at the dash-wheel:

1. Wince for half an hour in boiling bran-water;

2. Wince for half an hour or more in a dilute solution of chloride of soda or chloride of lime;

3. Boil the goods in soap-water containing half a pound of soap per piece;

4. Wince a second time in chloride of soda or lime, weaker than before;

5. Boil a second time in soap-water.

When the goods are dyed with Dutch madder and sumach,* soap cannot be well employed in the clearing process, but only bran and chloride of lime or chloride of soda, the latter being preferred. If dyed with the form of madder called *garancine* (see page 252), neither chloride of lime nor chloride of soda is admissible. Other precautions necessary to be attended to in the process of clearing and the finishing operations to which the calico is afterwards

No. 4.

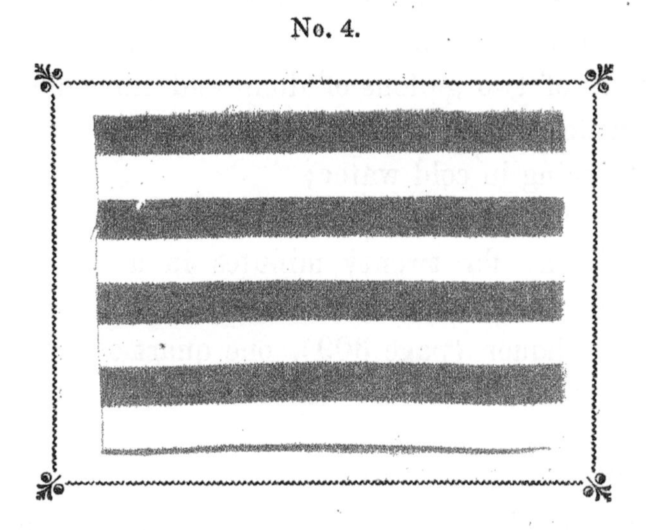

subjected have been noticed in the preceding section. The specimen, No. 4, is the finished cloth.

The strength of the mordant, solution of dung substitute, dye-beck, &c., and the details of the process generally, vary considerably for the same colour-

* The tints of cloths dyed in the madder-beck are considerably heightened by the addition of a small quantity of sumach. One pound of sumach with eighteen or twenty pounds of Dutch madder will dye as much stuff as twenty-four pounds of madder without sumach. The addition of astringent substances to logwood, peachwood, and cochineal, produces a similar result.

ing matter and mordant, according to the quantity or fullness of the figure, and the depth of its colour. To impart such a stripe as that in the preceding specimen, the bleached cloth may be submitted to the following operations:

1. Printing on mordant of red liquor (of spec. grav. 1·042, thickened with a pound and a half of flour to the gallon), and drying by being drawn over steam boxes;

2. Ageing for three days;

3. Dunging, 1°, in a mixture of four gallons of dung and three hundred of hot water; and 2°, in a mixture of two gallons of dung and three hundred of hot water;

4. Wincing in cold water;

5. Washing at the dash-wheel;

6. Wincing for twenty minutes in a solution of dung substitute and size, made of two quarts of substitute liquor (page 309), one quart of cleansing liquor (page 308), and three hundred gallons of water.

7. Wincing in cold water;

8. Dyeing in the madder-beck, containing about two pounds of madder per piece of twenty-eight yards;

9. Wincing in cold water;

10. Washing at the dash-wheel;

11. Wincing in soap-water, to which some perchloride or nitromuriate of tin has been added;

12. Washing at dash-wheel;

13. Wincing a second time in soap-water;

14. Wincing in a solution of bleaching powder of spec. grav. 1015 (3° Twad.)

15. Washing at the dash-wheel;
16. Drying by the "water extracter" (page 323);
17. Folding;
18. Starching (page 324);
19. Passing through the steam drying machine (page 325).

As the quantity of colouring matter which is deposited on the cloth in the dye-beck is much more dependent on the quantity of fixed mordant on the cloth, than on the strength of the dyeing liquid, a pattern comprising two or more different shades of the same kind of colour may be obtained by the same dye-beck, the cloth having been previously printed with the same kind of mordant at different strengths.

The pattern annexed (No. 5) is produced by the

No. 5.

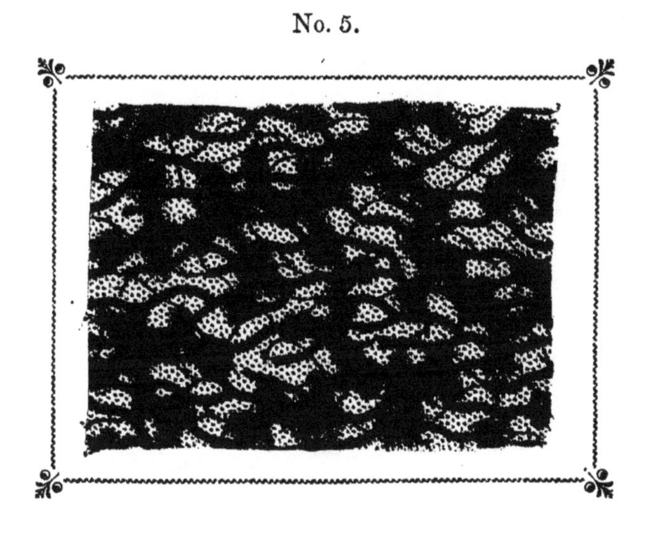

iron liquor mordant (pyrolignite of iron) at two different strengths, with the same infusion of madder. The mordant for the lilac is iron liquor of spec.

grav. about 1·010, thickened with three pounds and a half of British gum to the gallon. The mordant for the dark purple or black is iron liquor of spec. grav. about 1·020, thickened with a pound and a half of flour to the gallon. The two mordants may be printed on at once by the two-colour machine. The finest madder purples are obtained by a mixture of iron liquor with from five to six measures of "patent purple liquor," or "assistant mordant," (see page 294,) thickened with British gum.

To produce different shades of red on the same piece from the same madder-beck, the cloth may be printed with red liquor of any density between 3° and 25° Twaddell. The thickener usually employed for red liquor is either flour, starch, or a mixture of equal parts of flour and starch. The proportion of the thickening ingredient is varied according to the

No. 6.

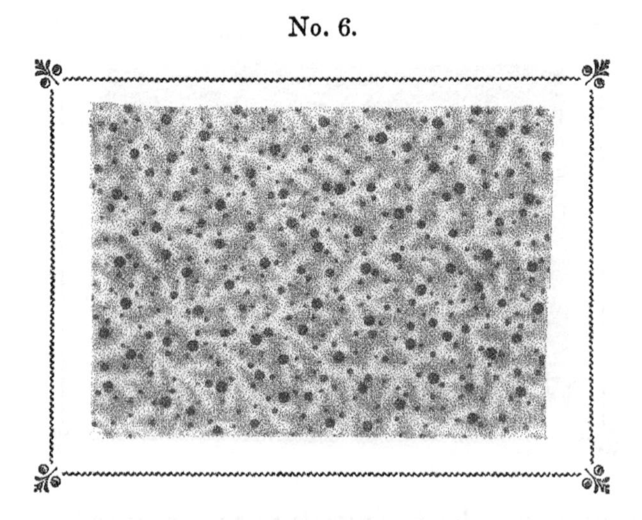

density of the mordant and the fullness or delicacy of the design; on the average, a pound and a half may be taken for a gallon of liquid. For a weak liquor and for a delicate figure, either British gum or gum arabic is substituted for starch or flour.

The mordants printed on the cloth for the specimen No. 6 were red liquor of spec. grav. 1·105 for the dark red spots, and the same liquid of spec. grav. 1·021 for the light red figure. Both solutions were thickened with four pounds of British gum to the gallon.

British gum is the best thickener for a solution of aluminate of potash.

As different mordants form compounds of very different colours with the same dye-stuff, a variety of colours may be communicated to the calico from the same infusion of colouring matter; provided as many kinds of mordants, or mixtures of mordants,

No. 7.

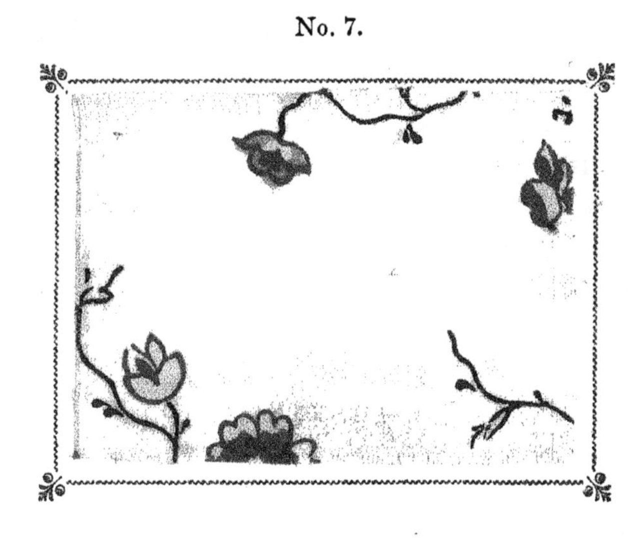

had been previously applied. The annexed pattern (No. 7), containing a design in black, purple, and two shades of red, was obtained by dyeing the cloth at one operation after it had been printed by the four-colour machine with the following mordants: iron liquor of spec. grav. 1·020 for the black; iron liquor of spec. grav. 1·012 for the purple; red liquor of spec. grav. 1·042 for the dark red; and red liquor of spec. grav. 1·010 for the pale red. The iron liquor for the black was thickened with a pound and a half of flour to the gallon. The weaker iron liquor for the purple and the two red liquors were thickened with three pounds of British gum to the gallon. After being printed, the cloth was "aged" three days, dunged, dyed with madder, soaped, and cleared in the ordinary manner.

To obtain on the same cloth the finest madder reds, purple, and black, it is sometimes better first to print on only the aluminous mordants for the reds, by the two or three colour machine, and then to age, dung, and madder. The strong iron liquor for black, and the weaker iron liquor for purple, may be next grounded in their proper places by hand-blocks, after which the drying, dunging, and maddering are repeated. Sometimes the mordants are printed on at different operations, but the dyeing is performed in one beck. For example, the mordant for black is printed on first by the single-colour machine, after which the cloth is aged for a day or two; the mordants for the other colours are then grounded in by the hand-blocks, and the ageing, dunging, dyeing, &c. are performed in the usual manner. An endless variety of tints, from red to chocolate, may be obtained from the same madder-beck by mixtures of the iron and aluminous mordant in different proportions.

Quercitron is a dyeing material well adapted for the madder style of work. With a mordant of red liquor of spec. grav. 8° or 12° Twaddell, thickened with starch, it affords a bright yellow; with iron liquor of spec. grav. 2° or 3° Twad., thickened with starch, an olive-grey colour; and with a mixture of the iron and aluminous mordants, a great variety of yellowish-olive tints. To produce a yellow ground with quercitron, the cloth may be padded in red liquor of 10° Twad., and after being dried, aged for two days, and winced in warm chalky water, be dyed in an infusion of quercitron (made of from two to three pounds per piece), containing some glue or size. To get a yellowish-olive figure from the same infusion, the cloth may be printed with a mixture of red liquor at 11°, and iron liquor at 5° Twad., in equal measures, and then dried, aged, dunged, winced in chalky water and dyed.

A very good orange is sometimes communicated to cotton goods in this style of work by dyeing in a mixed infusion of madder and quercitron, an aluminous mordant having been previously applied to the cloth. For a ground, the cloth may be padded in red liquor of 10° or 12° Twad., then winced in warm chalky water, and dyed in a decoction of two pounds of quercitron and a pound and a half of madder per piece. By varying the proportions of the madder and quercitron various shades of orange (from golden-yellow to scarlet) may be produced. An endless variety of cinnamon, olive, and fawn coloured tints may be obtained by applying to the cloth mixtures, in various proportions, of red liquor and iron liquor, and by dyeing the cloth in a mixed infusion of madder and quercitron. It is advisable not to have the temperature of the dyeing liquid above 140° or 150° Fahr.

Patterns in black and various shades of violet and purple may be imparted to cotton cloth in the madder style, by means of a decoction of logwood as the dye-stuff, with iron liquor and red liquor as the mordants. To produce a black ground, the cloth may be padded in a mixture of equal measures of red liquor at 8° Twad., and iron liquor at 6° Twad.; and after having been dried, aged, and winced in chalky water, it may be dyed in a decoction of logwood made from two pounds and a half or three pounds per piece, with the addition of a small quantity of sumach. A grey colour is obtained in the same way by using very weak iron liquor and a weak decoction of the colouring matter; and a violet colour, by applying weak red liquor to the cloth.

Cochineal is another dye-stuff, the colouring matter of which is capable of being imparted to cloth by the madder style. It is chiefly applied in this way as a ground on which figures are afterwards produced by other styles. To obtain an amaranth-coloured ground, the cloth is padded in red liquor of spec. grav. from 11° to 13° Twad., and after being dried, aged, and winced in chalky water, it is dyed in a mixed decoction of cochineal, galls, and bran. A beautiful orange is obtained by a mixture of decoction of cochineal and decoction of quercitron, with an aluminous mordant; and fine lilacs and violets by decoctions of logwood and of cochineal, with the same mordant.

The madder style admits of the application to cloth of the colouring matters of several dye-stuffs besides those which have been alluded to. By combining two or more in the same dyeing liquid, and by varying the mordants, an endless variety of tints may be obtained; but a detailed account of such processes could not be included in the limited plan of the present article.

II. TOPICAL AND STEAM COLOURS.

The mere mechanical part of the process is simpler in this than in any other style of calico-printing. The thickened solution of the colouring matter is applied topically (mixed with the mordant when any is required), in a state fit to penetrate to the interior of the fibre, and after the cloth is dried, the colouring matter is fixed, either by exposure to the air or a precipitating agent, or by the action of steam. The operations of dunging, dyeing, and clearing are here altogether omitted.

The vegetable colours which may be applied to cotton by this style of work are chiefly those which are not at all, or incompletely, precipitated from their solutions when mixed with a mordant; the deposition of the insoluble compound of mordant and colouring matter is usually determined in such cases by the surface attraction of the tissue, frequently assisted by the application of heat. Not many colours applied in this way are remarkable for their permanency. A few topical colours are printed on the cloth in an insoluble state, as pigments, and remain attached principally through the starch or gum used as thickening.

The topical colours most extensively used in printing cottons are the following:

Black.—A very good topical black, known as "chemic black," may be imparted to cotton by means of a mixture of decoction of logwood, copperas, and pernitrate of iron, in the proportions of either of the recipes following:

No. 1.

1 gallon of logwood liquor (decoction of logwood) of spec. grav. from 6° Tw. to 8° Tw.,
4 ounces of copperas,
$\frac{1}{8}$ to $\frac{1}{12}$ gal. (according to the strength of the logwood liquor) of solution of pernitrate of iron of spec. grav. 50° Tw.
Thicken with either flour or starch.

No. 2.

1 gal. of logwood liquor at 8° Tw.,
2 ounces of copperas,
1 pint of solution of pernitrate of iron of 8° Tw.,
1$\frac{1}{4}$lbs. of starch.

The logwood liquor, copperas, and starch are boiled together for a few minutes, and when the mixture is cooled to about blood-heat the nitrate of iron is added.

This mixture may be printed on the white calico by the roller at the same time as the mordants for colours to be afterwards applied by the madder style.

After ageing for two days the black colour becomes so permanently attached to the cloth that it is very little affected by the remaining operations of dunging, washing, dyeing, and clearing.

Brown.—A very fast topical brown, forming the ground in the annexed specimen (No. 8), may be

No. 8.

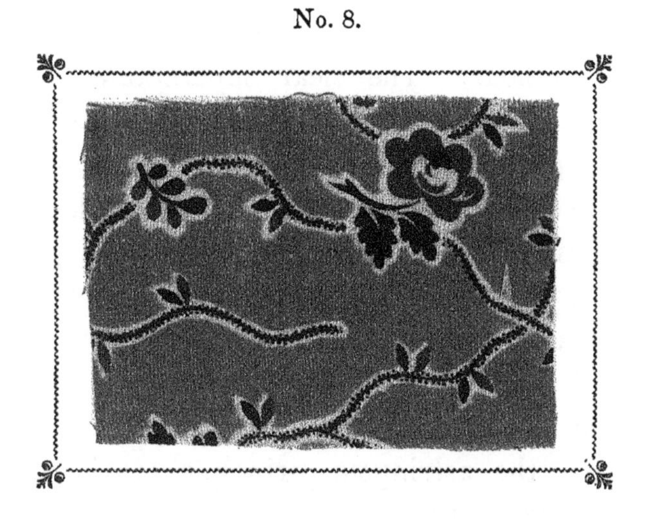

communicated to cotton goods by a solution of catechu, mixed with a salt of copper and muriate of ammonia. The following proportions have been recommended to me by an eminent Lancashire printer:

1½ pounds of catechu,
¾ to 1 gill of solution of nitrate of copper of 90° Tw.,
6 to 8 ounces of muriate of ammonia,
1 gallon of water.
Thicken with either gum senegal or British gum.

Some printers are in the habit of using a mixture containing much less muriate of ammonia and more nitrate of copper than the above, such as the following:

1½ pounds of catechu,
3 ounces of muriate of ammonia,
2 quarts of water,
1 quart of pyroligneous acid of 2° Tw.,
3 pounds of gum senegal, and about
½ pint of solution of nitrate of copper of 100° Tw.

The catechu and muriate of ammonia are first mixed with the water and pyroligneous acid; the mixture is then heated and kept at the boiling point for ten minutes, after which it is allowed to settle for half an hour. The clear supernatant liquid is then decanted from the insoluble matters and mixed with the gum senegal, and when the latter is dissolved the solution of nitrate of copper is added.

The first of the preceding preparations probably deserves a preference. Both of them, like the above topical black, may be printed on the cloth by the compound machine, with mordants for colours to be applied by the madder style. During the ageing of the goods, the astringent principle of the catechu becomes fixed on the cloth in an insoluble state, but it is usual to complete the fixation of the colouring matter by passing the cloth through a solution of bichromate of potash previous to being dyed in the madder-beck. The specimen No. 8 is an example of a design in madder colours on the catechu ground; the manner of producing such a pattern will be again adverted to, as a part of the process is referable to the resist style of work.

The chemical change which takes place during the fixation of the astringent principle of the catechu is the absorption of oxygen. Catechu, or its colouring principle, exists in two forms; deoxidized and soluble in water, and oxidized and insoluble in pure water, but soluble in an aqueous solution of the deoxidized catechu. The form in which catechu is applied to cloth in the above mixtures is as a solution of the oxidized in that of the deoxidized catechu; and the chloride of copper, formed by the reaction of muriate of ammonia on nitrate of copper, acts as a slow oxidizing agent (through the decomposition of water) to the deoxidized catechu.

Spirit purple.—For a spirit or fancy* purple, such as that in the following specimen (No. 9), a de-

No. 9.

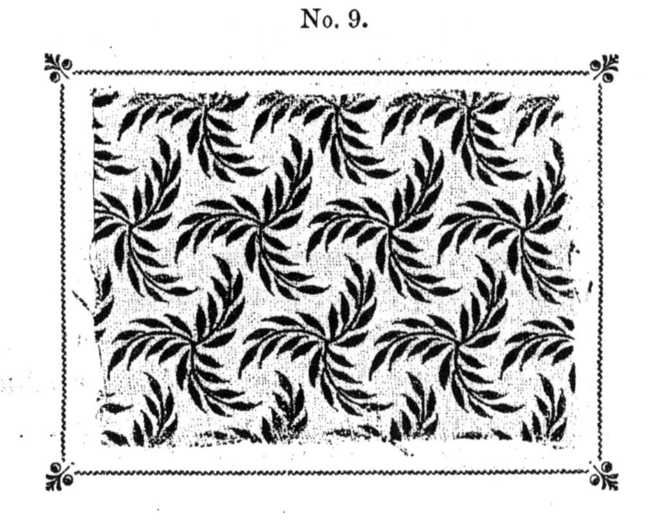

coction of logwood, mixed with a tin or an aluminous mordant, is commonly employed. A gallon of logwood liquor of 6° Tw. may be boiled for a few minutes with a pound of starch, and when the mixture is lukewarm, there are added, first, a pint and quarter of solution of perchloride (nitromuriate) of tin at 120° Tw., and afterwards, a quarter of a pint of oil.

* All fugitive topical colours not fixed by steaming are termed spirit, fancy, or wash-off colours.

This mixture should be carefully stirred before being applied to the cloth. Some printers use a little less tin than is prescribed above, and add a very small quantity of pernitrate of iron.

After being printed with this mixture, the cloth should be suspended in a warm room for two days and two nights, and then washed at the rinsing machine.

Spirit chocolate.—A fancy chocolate or puce colour may be imparted to cotton by means of a mixture of logwood liquor, peachwood liquor, perchloride of tin, and a little nitrate of copper, thickened with either starch or British gum.

A much faster colour may be obtained by first printing on a mixture of logwood liquor, red liquor, and oxalic acid, and after ageing, passing the cloth through a solution of bichromate of potash. If the materials are employed sufficiently strong, an excellent black may be imparted by such a process. The black or chocolate coloured substance thus produced is a compound of oxide of chromium with the colouring matter of logwood.

Spirit pink.—The following receipt for a topical pink has been recommended to me by a Lancashire printer:

1 gallon of peachwood liquor at 8° Tw.,
1¼ pounds of starch,
½ gill of solution of nitrate of copper at 100° Tw.,
3 gills of solution of perchloride of tin,
4 ounces of "pink salt" (see page 289), and
1 gill of oil.

The peachwood liquor is first boiled briskly with

the starch; the nitrate of copper is next added; and when the mixture is cooled to about blood-heat, the remaining ingredients are introduced.

Spirit yellow.—A fancy yellow, pretty much employed for grounding in by the block, is a mixture of decoction of French berries, red liquor, and oxalic acid, thickened with starch or gum tragacanth. Some printers use, instead of red liquor and oxalic acid, a mixture of perchloride of tin and alum, and thicken with starch; others, perchloride of tin only, with gum as the thickener; and others, a mixture of red liquor, alum, and protochloride of tin, thickened with either flour or starch.

Blue.—1. The following topical blue is sometimes grounded in by the block, for light goods, after all or most of the other colours have been applied. Being easily washed out, it is not used except with fugitive colours, and in cases where it would be inconvenient to apply the mixture for steam blue.

> 1 pound of yellow prussiate of potash,
> 1½ gills of solution of nitrate of iron of 80° Tw.,
> 3 gills of perchloride of tin of 100°,
> 1 gallon of water, and
> 1¼ pounds of starch.

The prussiate of potash is first dissolved in the water, and with this solution the starch is boiled briskly for a few minutes. When the mixture is cooled, the nitrate of iron and perchloride of tin are added. The blue colour is derived from Prussian blue (see page 256).

2. Indigo communicates a faster blue, so far as

the action of soap and alkalies is concerned, than Prussian blue; but the latter possesses considerably more brilliancy than the former in its ordinary state. As a topical colour, indigo is applied in the form of indigotin, or the hydruret of indigo-blue (see page 244); the deoxidizing agent employed to produce the indigotin being either metallic tin, or the red sulphuret of arsenic (red orpiment, or red arsenic). Until within a few years, almost the only solution of indigo employed as a topical colour was that known as "pencil blue;" which is prepared by mixing with water and boiling together, about equal parts of ground indigo, orpiment, and quick-lime, and when the mixture is withdrawn from the fire and become lukewarm, adding about as much carbonate of soda as of orpiment previously introduced. The clear liquor when thickened with gum was applied to the cloth by a pencil, or by the block charged with the colour by a particular contrivance to prevent as much as possible the access of air to the sieve.

3. A much more convenient way of effecting the conversion of indigo-blue to white indigo, through the action of the same deoxidizing agent, is to mix the ground indigo with a solution (previously made) of red arsenic in perfectly caustic potash or soda, adding as much more caustic alkali as is necessary to keep the white indigo in a state of solution. In this way, the inconvenience arising from the sediment of lime and excess of orpiment is avoided.

Pencil blue contains a considerable excess of the deoxidizing agent, which is necessary on account of the rapidity with which indigo-blue is deposited from all solutions of indigotin when freely exposed to the air. With a smaller proportion, the indigo-blue might be deposited in an insoluble state on the surface of the cloth before sufficient time is allowed for the solution of indigotin to penetrate to the interior of the fibre. This inconvenience may be partly surmounted by directing a jet of coal-gas against the printing roller, and a short length of the cloth passing from the roller.

The construction of the cylinder machine, by which pencil blue is applied, differs slightly from that represented in figure 39, page 336. The cylinder *c* is dispensed with, the engraved roller itself dipping an inch or two into the colour mixture; and the roller is cleaned from the superfluous colour by revolving in close contact with one of the sides of the colour-troughs, which thus acts as the "doctor."

4. The deoxidizing agent employed for reducing or affording hydrogen to the indigo in the preparation now commonly substituted for pencil blue, is metallic tin. One equivalent of the metal becomes oxidized, in the presence of a caustic alkali, at the expense of two equivalents of water, the hydrogen of which unites with the indigo. To prepare such a mixture fit for application by the roller, the materials may be employed in the following proportions:

4 pounds of ground indigo,
4 quarts of water,
2½ quarts of solution of caustic soda of 70° Tw.,
3 pounds of granulated tin.

The indigo is first intimately mixed with the water, the tin and alkali are afterwards added, and the mixture is heated to the boiling point, then taken off and stirred until a drop, placed on a glass plate, appears of an orange-yellow colour. To this solution is afterwards added a mixture of a solution of chloride of tin at 120° Tw. with an equal measure of muriatic acid, until the free alkali is completely neutralized, and an olive-coloured precipitate falls. The mixture is then well stirred, and added to strong gum-water to the required shade.

For some purposes, the free alkali is neutralized by tartaric acid, and from half a gill to a gill of the solution of tin is afterwards added.

It is evident that the above preparation contains white indigo in an insoluble state; in a form, therefore, unable to enter the interior of the fibres. To dissolve the white indigo and allow it to be absorbed, the printed cloth is passed through a solution of carbonate of soda of spec. grav. 8° or 10° Tw., and at the temperature 80° or 90° Fahr. On afterwards exposing the cloth to the air, the solution of white indigo, now within the fibres, absorbs oxygen and affords a precipitate of indigo-blue.

Steam colours.—Very few colours attach themselves firmly to the cloth by being merely printed on together with the mordant; but if a cloth thus printed is exposed for a short time to the action of steam, an intimate combination takes place between the tissue, colouring matter, and mordant. Before the printed cloth is exposed to steam, the colouring matter may in general be easily removed by washing with pure water; but afterwards it is attached to the tissues almost as strongly as in any other style of printing, presenting, moreover, a brilliancy and delicacy hardly attainable by any other process. Printing by steam is one of the most important of modern improvements in calico-printing; it is practised not only on goods of cotton, but also on silk, woollen cloths, and chalys.

The brilliancy and permanency of almost all steam colours are greatly increased by impregnating the cloth with a solution of tin, or, for some styles, with a solution of acetate of alumina, previous to the application of the colours. The solution of tin now commonly used for this purpose is the *stannate of potash*, which is, when properly made, a solution of peroxide of tin in caustic potash (see page 289); this preparation sometimes contains protoxide of tin, but the stannate containing the peroxide only is preferred. This alkaline solution is not nearly so injurious to the cotton fibre as the perchloride.

After having been padded in the solution of stannate of potash, the pieces of cotton are usually passed through a cistern containing a solution of muriate of ammonia, to produce a precipitate of peroxide of tin. Some printers employ very dilute sulphuric acid instead of a solution of muriate of ammonia, but the latter is decidedly preferable.

To the cloth thus prepared, or occasionally without any preparation except bleaching, the solutions of the mixed colouring materials and mordants, properly thickened, are applied either by the roller or block. Steam colours are chiefly grounded in by the block to cloths which have been already printed and finished off according to other styles of work, particularly the madder style. The following recipes will afford examples of the principal mixtures which are applied to cotton as steam colours; some of them may also be applied to silk and woollen goods, but for this purpose the proportions of the materials generally require to be varied. The mordant most frequently used for steam colours is red liquor, mixed with oxalic or some other acid to prevent the precipitation of the compound of colouring matter and mordant.

Steam black.—The first of the mixtures following is best adapted for the roller, the other for grounding in by the block:

No. 1.

1 pint of red liquor of 18° Tw.,
2 pints of iron liquor of 24° Tw.,
1 gallon of logwood liquor of 8° Tw.,
1¾ pounds of starch,
1½ pints of pyroligneous acid of 7° Tw.

All these materials may be mixed promiscuously and then boiled for a few minutes to form a mucilage. The cotton requires to be steamed about thirty minutes.

No. 2.

3½ pints of peachwood liquor of 6° Tw.,
7 pints of logwood liquor of 6° Tw.,
12 ounces of starch,
14 ounces of British gum,
3 ounces of sulphate of copper,
1 ounce of copperas,
3 ounces of a neutral solution of pernitrate of iron, made by mixing one pound of acetate of lead with three pounds of the common acid nitrate of iron of 122° Tw.

If intended for goods of silk and wool, four ounces of extract of indigo should be added.

The logwood liquor and peachwood liquor are mixed and divided into two equal portions, one of which is boiled for a short time with the starch, and the other with the British gum. The two liquids are afterwards mixed, and the remaining ingredients are added; the nitrate of iron being introduced last, and not before the mixture is cold.

Steam red.—The best steam red for cotton is obtained by decoction of cochineal, with oxalic acid and protochloride of tin. The mixture obtained according to the following receipt may be applied either by the roller or block:

1 gallon of cochineal liquor of 6° Tw.,
1 pound of starch,
3 ounces of oxalic acid,
4 ounces of cryst. protochloride of tin.

The cochineal liquor is first boiled with the starch for a few minutes; when the mixture is half cold, the oxalic acid is added, and as soon as the acid is dissolved the salt of tin is introduced.

A cheaper but less brilliant steam red, much used by some printers, is prepared by substituting peachwood liquor for cochineal liquor in the above.

Steam purple.—To a gallon of red liquor of 18° Tw., heated to about 140° Fahr., three pounds of ground logwood are added; the mixture is well stirred for about half an hour, and then strained through a cloth filter, the residue on the filter being washed with two quarts of hot water, which are received into the first liquid. The mixture thus obtained may be diluted with water, according to the shade of colour required; for a moderate depth, one measure may be mixed with three of water, and thickened with starch, flour, or gum. This preparation may be applied either by block or roller.

Steam yellow.—Either decoction of Persian berries, decoction of quercitron, or decoction of fustic, may be used as a steam yellow, but the first is most commonly employed at present.

No. 1.

1 gallon of berry liquor of 4° Tw.,
5 ounces of alum, thickened with about
14 ounces of starch.

No. 2.

1 gallon of berry liquor of 4° Tw.,
1½ gill of red liquor of 18° Tw.,
2 ounces of crystals of protochloride of tin, and about
14 ounces of starch.

The mixture made according to the following receipt affords a darker shade than either of the preceding:

No. 3.

1 gallon of a mixture of equal measures of decoction of Persian berries at 15° Tw., and of decoction of fustic at 15° Tw.,
14 ounces of starch,
7 ounces of alum,
7 ounces of crystals of protochloride of tin.

The decoctions of the dye-stuffs are mixed with the alum and starch, and heated until properly thickened; the mixture should be soon withdrawn from the fire, and when cold mixed with the salt of tin.

The preparation made as No. 2 will probably be found superior to either of the others for cotton goods. The steaming for No. 3 must be continued only a short time, else the fibre of the cotton would be apt to become corroded by the salt of tin. This preparation is better adapted (as a steam colour) for fabrics of wool and silk than for those of cotton, but it may be advantageously applied to cotton as a spirit or wash-off colour (page 363).

The orange stripe in the specimen annexed (No.

No. 10.

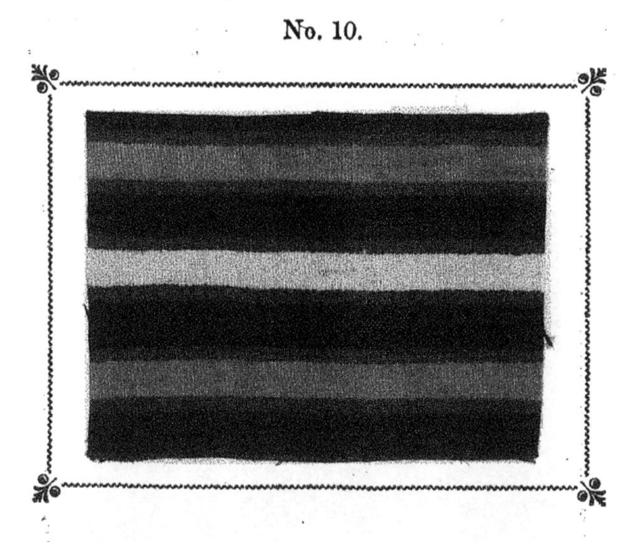

10) is also produced by decoction of Persian berries, the mordant being protoxide of tin only.

A convenient mixture for producing this colour is the following:

1 gallon of berry liquor made from three pounds of berries to the gallon, and
4 ounces of cryst. protochloride of tin. Boil together for a few minutes and thicken with
3 to 4 pounds of British gum or 1 pound of starch.

The cloth may be steamed and washed in the usual manner, but this colour becomes strongly attached by merely ageing the cloth for two or three days, and then passing it through hot chalky water.

Steam blue.—A very beautiful steam blue may be communicated to cotton and woollen goods by means of a mixture of yellow or red prussiate of potash, with tartaric, oxalic, or sulphuric acid, and alum or perchloride of tin. If for applying to cotton goods, alum is used; but if for woollen fabrics, perchloride of tin is preferable. The blue in the annexed specimen (No. 11) was produced by such a process.

No. 11.

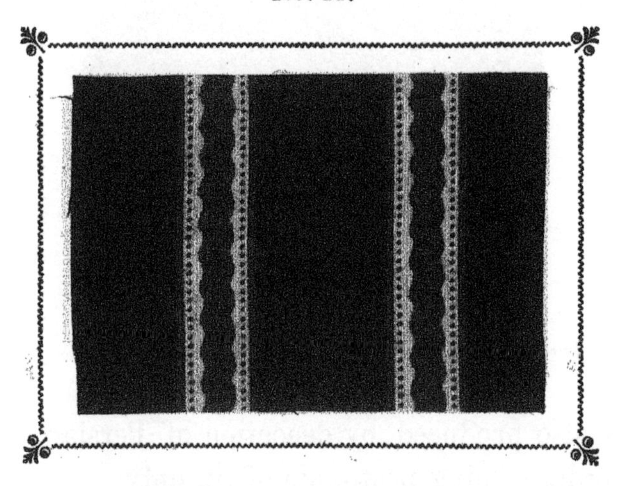

For printing on cottons by the roller, either No. 1 or No. 2 of the following mixtures may be used:

No. 1.

1 gallon of water,
1¼ pounds of yellow prussiate of potash,
3 to 4 ounces of alum,
5 to 6 ounces of oil of vitriol,
1½ pounds of starch.

No. 2.

1 gallon of water,
1¼ pounds of yellow prussiate of potash,
3 to 4 ounces of alum,
10 to 12 ounces of tartaric acid,
1½ pounds of starch.

The starch and prussiate of potash are boiled in the water, and when the mixture is withdrawn from the fire and cooled, the sulphuric or tartaric acid and alum are introduced. The mixture made as No. 2 affords a more lively colour than that made as No. 1, but the latter is least expensive.

No. 3.

1 gallon of water,
3 to 3½ ounces of alum,
1½ to 2 ounces of oxalic acid,
3 to 4 ounces of tartaric acid,
20 ounces of gum,
12 ounces of yellow prussiate of potash.

The gum, acids, and alum may be first dissolved in the water with the assistance of heat, and when the mixture is quite cold, the prussiate of potash is added.

The time necessary for steaming cottons printed with either of these preparations is about thirty minutes. When withdrawn from the steaming cylinder or chamber, the goods present, if *yellow* prussiate of potash is used, a blueish-white colour, which changes to deep blue on exposure to the air for a couple of days. The chemical change by which the colour is produced during the exposure to air depends on the absorption of oxygen or the removal of hydrogen; as is evident from the circumstance, that if the goods are passed through a solution of bichromate of potash as soon as withdrawn from the steaming cylinder or chamber, the blueish-white changes to deep blue immediately. If the *red* prussiate of potash is employed instead of the yellow prussiate, the cloths acquire the proper blue colour during the steaming, and the depth of the colour is not sensibly increased by exposure to air or to a solution of bichromate of potash.

The blue colouring matter produced in these processes is a variety of Prussian blue, formed through the decomposition of hydroferrocyanic acid set at liberty by the action of the more powerful acids present on the prussiate of potash. When an aqueous solution of pure hydroferrocyanic acid is gently heated and exposed to the air, the acid suffers decomposition with formation of hydrocyanic or prussic acid, and Prussian blue, which precipitates. Assuming the composition of the Prussian blue thus formed to be the same as that produced by mixing a solution of yellow prussiate of potash with a solution of a salt of the peroxide of iron, (which contains cyanogen and iron in the proportion of nine equivalents of the former to seven equivalents of the latter,*) the decomposition which the hydroferrocyanic acid experiences appears to be after the following manner: seven equivalents of the acid, containing twenty-one equivalents of cyanogen, seven equivalents of iron, and fourteen equivalents of hydrogen, afford, 1°, one equivalent of Prussian blue, containing seven equivalents of iron and nine equivalents of cyanogen; 2°, twelve equivalents of hydrocyanic acid; and 3°, two equivalents of hydrogen to be removed by an oxidizing agent. This decomposition may be expressed more simply in symbols; thus,

* The prussiate of potash which enters into the composition of this variety of Prussian blue may be neglected in the above calculation.

7 eqs. of hydroferrocyanic acid ($H_2 + Fe\,Cy_3$) or	H_{14}	Fe_7	Cy_{21} produce
1 eq. Prussian blue		Fe_7	Cy_9
12 eqs. hydrocyanic acid	H_{12}		Cy_{12}
2 eqs. hydrogen	H_2		

Such we may suppose to be the decomposition which takes place when an aqueous solution of hydroferrocyanic acid is heated with exposure to the air. That the reactions which occur in the steam blue process are somewhat different from the above, however, is pretty evident from the circumstance that the colour does not appear until the cloth is exposed to a source of oxygen, although the acid is certainly decomposed during steaming, as is manifest from the odour of hydrocyanic acid then developed. According to another, and a more consistent, view of these changes, the blueish-white compound on the steamed cloth, before being exposed to the air, is the same as the precipitate which falls on mixing a solution of yellow prussiate of potash with a solution of a salt of the protoxide of iron. This precipitate contains, besides a certain proportion of prussiate of potash, iron and cyanogen in an equal number of equivalents. When exposed to the air, it absorbs oxygen and becomes deep blue; the oxygen thus absorbed combines with a portion of the iron in the precipitate, forming peroxide of iron, which remains as an essential part of the Prussian blue.

As the blueish-white precipitate contains a compound of an equal number of equivalents of iron and cyanogen united with prussiate of potash, it may be formed from a mixture of prussiate of potash and hydroferrocyanic acid, with the separation of nothing more than hydrocyanic acid; thus,

From 9 eqs. of hydroferrocyanic acid	=	H_{18}	Fe_9	Cy_{27}
Deduct 18 eqs. of hydrocyanic acid	=	H_{18}		Cy_{18}
There remains			Fe_9	Cy_9

On exposure to the air, this compound of iron and cyanogen, the probable constitution of which is represented by the formula $Fe_6 + 3(Fe\ Cy_3)$ absorbs three equivalents of oxygen, and thereby affords a compound of peroxide of iron with the variety of Prussian blue noticed in the preceding page. The formula for this compound is $Fe_2\ O_3 + (Fe_4 + 3(Fe\ Cy_3))$.

The reactions which occur when the red prussiate of potash is employed are different, the acid liberated from that salt by the action of the stronger acids not having the same composition as hydroferrocyanic acid. The composition of hydroferridcyanic acid (the acid liberated from the red prussiate) is such as to allow of the decomposition of the acid into Prussian blue, hydrocyanic acid and cyanogen, without the interference of atmospheric air or any other source of free oxygen.

In its present form, this beautiful colour has not been long in general use for application to calicos. The colour obtained by the mixture formerly employed, consisting of prussiate of potash with tartaric or sulphuric acid, without any addition of perchloride of tin or alum, is always lighter in shade and less vivid than that obtained with such an addition, however concentrated the solution of prussiate of potash. The acids in the mixture, including the sulphuric acid of the alum in combination with alumina (namely three equivalents for one equivalent of alum), should be in sufficient quantity to neutralize one equivalent of alkali for every two-thirds of an equivalent of prussiate of potash; or to saturate 5·9 ounces of anhydrous potash for 18 ounces of the prussiate.

Steam green.—A very good steam green may be communicated to cotton goods by combining the materials for producing a yellow, with the preceding mixture for steam blue; thus,

1 gallon of berry liquor made from a pound and a half of Persian berries (or of 4° Tw.),
12 ounces of yellow prussiate of potash,
3 to 4 ounces of crystals of protochloride of tin,
5 to 6 ounces of alum,
3 to 4 ounces of oxalic acid.
Thicken with gum.

The oxalic acid, the muriatic acid derived from the salt of tin, and the sulphuric acid united with alumina in the alum, should form, together, one equivalent, or a quantity sufficient for the saturation of one equivalent of a protoxide for every two-thirds of an equivalent of the prussiate. The time required for steaming this colour is about thirty minutes.

After the colour mixtures are printed on, the calico is dried in a warm atmosphere for two days or thereabouts before being exposed to the action of the steam. Different methods of applying the steam are practised in different print-works. In some the goods are introduced into a large stout deal box, the lid of which is made very nearly steam-tight by edges of felt. The steam is admitted near the bottom by a thickly perforated pipe which traverses the box. For the deal box is sometimes substituted a small chamber built of masonry, about four or five feet in length, by three feet in width, and three feet in height. The cloth is suspended free from folds, on strings across the chamber.

The most common method of applying the steam is the following. Three or four pieces of the printed and dried calico are stitched together at the ends and coiled round a hollow cylinder of copper, about three feet in length and four inches in diameter, and perforated with holes about one-twelfth of an inch in diameter and half an inch distant from each other. One of the ends of the cylinder is open, to admit the steam; the other is closed. The calico is prevented from coming immediately into contact with the cylinder by a roll of blanket stuff, and is covered with a piece of white calico tightly tied around the roll. During the lapping and unlapping of the calico the column is placed horizontally in a frame, in which it is made to revolve; but during the steaming it is fixed upright, and supplied with steam through its bottom from the main steam boiler of the works, the quantity admitted being regulated by a stopcock. During the whole process the temperature of the steam should be as near 211° or 212°, as possible: the condensation which takes place below that degree is apt to cause the colours to run; but a higher temperature is also injurious, as a slight condensation, sufficient to keep the goods always moist, is essential to the success of the process. The steaming is continued for from twenty minutes to three quarters of an hour, according to the nature of the stuff and the colouring mixture. The usual time with cottons is twenty-five minutes, and with dé laines about thirty or thirty-five minutes. The time required for steaming cotton goods by the chamber is longer than what is required by the column, being generally about an hour. When the steam is cut off, the cloths should be immediately unrolled to prevent any condensation: they are then soft and flaccid, the material used as a thickener for the colours being in a semi-fluid state; but on exposure to the air for a few seconds only, the thickener solidifies, and the goods become perfectly dry and stiff. After the pieces have been aged for a day or two, the thickener is separated by a gentle wash in cold water.

To produce with steam colours only such a pattern as the annexed (No. 12), containing a design in lilac, pink, red, yellow, black, and dark orange red, the cloth may be printed by the five-colour machine in the following manner:

No. 12

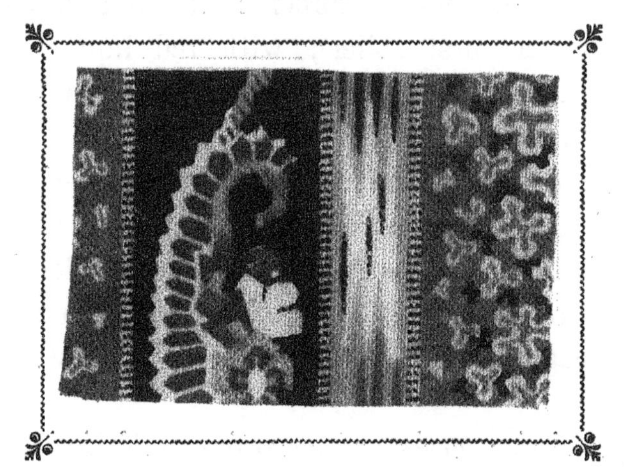

By the first roller, with a mixture of logwood liquor, starch, and solution of tin for producing the lilac;

By the second and third rollers, with the mixtures for the pink and red (see page 371), one containing weaker cochineal or peachwood liquor than the other;

By the fourth roller, with the mixture for the yellow (see page 372);

By the fifth roller, with the mixture for steam black (see page 370).

The dark orange red results from the mixture of the red with the yellow. After being steamed, the cloth is aged in a warm room for two days and two nights, and then washed at the rinsing machine.

The style following, for producing a design in black, red, brown, green, and yellow on a white ground, is a combination of the madder style with a topical brown and steam colours, which is susceptible of a great variety of interesting modifications.

1. The cloth is printed by the three-colour machine in the following manner: with iron liquor, for black, by the first roller; with red liquor by the second roller, and with catechu brown (page 361) by the third roller.

2. After being printed, the cloth is aged for two days, dunged, dyed in the madder-beck, and cleared.

3. The cloth is lastly printed by the block with the mixtures for steam green (page 379), and steam yellow (page 372), then steamed, aged, and washed.

By a similar series of operations, a design may be

imparted in black, brown, lilac, pink, green, blue, orange, and yellow, on a white ground. The cloth is first printed by the four-colour machine with iron liquor of two strengths, one for the black, the other for the lilac; with red liquor for the pink, and with the mixture for catechu brown (page 361). After being aged, dunged, dyed with madder, and cleared as usual, the cloth is printed by the block with the mixtures for steam blue (page 374), and steam yellow (page 372), and then steamed in the ordinary manner. To produce the orange, the steam yellow is printed on a part of the pink, and the green results from the mixture of some of the yellow with the blue.

The specimen No. 10, page 373, is an example of the combination of madder colours with steam colours; for the red and chocolate stripes, the cloth was printed with red liquor and the mixture of red liquor with iron liquor, and after dunging, dyeing, and clearing in the usual manner, the mixture for steam orange was applied by the block.

3. MINERAL COLOURING MATTERS.

Mineral colouring matters are adapted, not only to the production of designs on a white or coloured ground, but also to form a ground for the reception of a design in other colours. To impart the colour to the entire surface of the cloth, the latter may be impregnated successively, by the padding machine, with the two solutions necessary to produce the colour, or else the cloth may be padded in one of the solutions and be afterwards winced in the other.

To produce a design in a mineral colouring matter on a white or coloured ground, the cloth is usually first printed with one of the solutions, and then either padded or winced in the other.

The common "padding machine," by which a cloth is uniformly imbued with a liquid, is much the same as the starching machine (fig. 37, page 324), the thickened liquid being contained in the trough *a*.

A simpler padding machine is employed in some print-works, of which fig. 41 is a representation: *a* and *b* are two cylinders revolving in close contact; *b* is covered with blanket stuff, and dips partly into a trough which contains the liquid slightly thickened. The calico is first rolled around the cylinder *c*, and then passed in the direction of the arrow between the cylinders *a* and *b*. As soon as the calico is padded, it is dried hard, by exposure to a temperature of 212° or thereabouts, either by being drawn over a series of sheet-iron boxes heated by steam, or through the steam drying machine described at page 325, or through the hot-flue. If the colour is to be applied to the face of the cloth only, and not to both face and back, the common printing machine with a roughened roller is substituted for the padding machine.

Fig. 41.

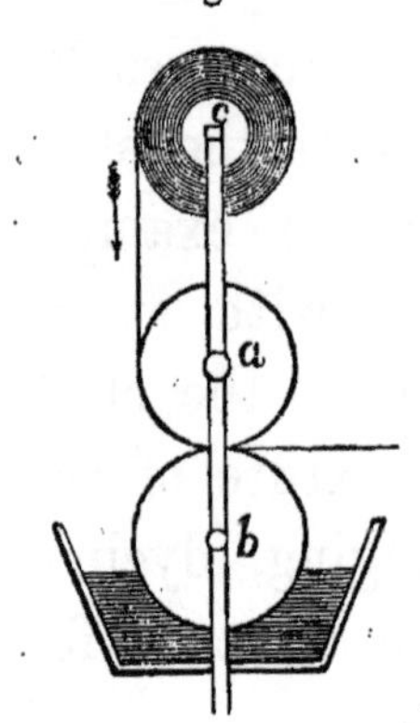

Prussian blue.—More than one method of applying this colouring material to cloth has already been noticed; another is by the consecutive application of either yellow prussiate of potash and a salt of the peroxide or protoxide of iron, or red prussiate of potash and a salt of the protoxide of iron. The latter method is very rarely practised. To impregnate the entire surface of a piece of cloth with Prussian blue, it may be treated in the following manner:

1°. Pad in a solution of acetate and sulphate of iron made by adding three pounds of acetate of lead to a solution of four pounds of copperas in a gallon of water, decanted from the precipitated sulphate of lead and diluted to the density 2° or 3° Tw.

2°. Dry the cloth, and then wince it in warm chalky water.

3°. Wince it in a solution of a pound of yellow prussiate of potash in forty gallons of warm water, to which are added four ounces of oil of vitriol.

To produce a design in Prussian blue by this style of work, the cloth may be printed with the mixed solution of acetate and sulphate of iron, made as above, of spec. grav. 4° or 5° Tw., thickened with gum and "sightened" by the addition of a little prussiate of potash. After being aged, the cloth is winced in chalky water, cleaned, and winced until it acquires the desired shade in a solution containing three or four ounces of prussiate of potash, and one fluid ounce of muriatic acid per piece.

Chrome-yellow.—The yellow and orange in the specimen annexed (No. 13) are produced by the two chromates of lead, chrome-yellow and chrome-orange.

To impart a ground of chrome-yellow to a piece of calico, the cloth should be padded with a solution of two pounds of acetate of lead in a gallon of water

No. 13.

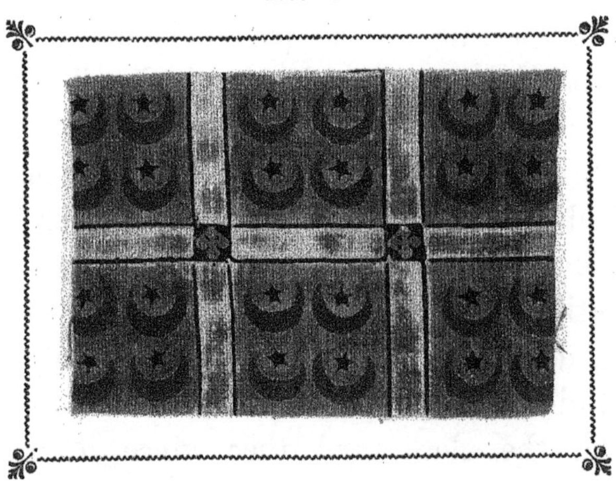

containing a little size, then dried, passed first through a weak solution of carbonate of soda, and afterwards through a solution of bichromate of potash. Rinse and dry.

To apply chrome-yellow topically, the cloth may be printed with a solution containing both acetate and nitrate of lead (from seven to ten ounces of each to the gallon) thickened with starch. After being printed and dried, the cloth is winced first in a weak solution of carbonate of soda, and afterwards in a solution of bichromate of potash containing about two ounces per piece. To clear the whites, the cloth may be winced in water slightly acidulated with muriatic acid.

Chrome-orange.—A ground of chrome-orange may be communicated to a piece of cotton by first applying chrome-yellow in the ordinary manner, and then exposing the cloth to boiling lime-water, which withdraws a portion of the chromic acid from the chrome-yellow and leaves chrome-orange: thus,

1. Pad the cloth twice in a saturated solution in water of acetate and nitrate of lead, in the proportion of a pound of the nitrate to a pound and a quarter of the acetate.* Dry in the hot-flue.

2. Wince in weak milk of lime for a few minutes.

3. Wince in a warm solution of bichromate of potash containing five or six ounces per piece; and lastly,

4. Wince in boiling milk of lime. Rinse and dry.

To produce a design in chrome-orange on a white ground, the cloth may be printed with a saturated solution of acetate and nitrate of lead (as above) thickened with British gum; after being dried, it is passed through a solution of sulphate of soda to fix the oxide of lead in an insoluble state, then well washed in water, and winced in a warm solution of bichromate of potash. It is afterwards rinsed, and passed through boiling milk of lime to convert the chrome-yellow into chrome-orange.

A design in chrome-yellow on a chrome-orange ground may be obtained by printing an acid on the orange ground, so as to withdraw the excess of oxide of lead from the subchromate (orange), and thus form the neutral chromate (yellow). The specimen No. 13 was produced in this way; the blue and black are merely spirit colours.

Different shades of green are given to cotton goods by a mixture of chrome-yellow with Prussian blue. The colouring materials may be applied consecutively, or at once from mixed solutions; the cloth being

* Water is capable of dissolving nearly twice as much of a mixture of acetate and nitrate of lead, in the proportion of single equivalents, as of either of the salts separately.

first padded, in the latter case, with a mixture of acetate of iron and nitrate of lead, and afterwards winced in a solution of prussiate of potash and bichromate of potash with a small quantity of muriatic acid.

To obtain a design of a green colour by conjoining chrome-yellow with indigo-blue, the cloth may be printed with a solution of from two pounds to two pounds and a half of nitrate of lead in a gallon of the neutralized mixture of white indigo with solution of tin prepared as described at pages 367, 368. After being printed, the cloth is passed, first, through a warm solution of carbonate of soda to fix the blue and oxide of lead, and afterwards through a solution of bichromate of potash to raise the yellow.

Iron buff.—The solutions of iron in common use for iron buff are the pernitrate and a mixture of the acetate with the protosulphate, obtained by adding from one part to three parts of acetate of lead (pyrolignite) to three parts of copperas. Double decomposition takes place between the acetate of lead and a portion of the copperas with formation of acetate of iron and sulphate of lead. For producing light shades, alum is sometimes added, together with a little carbonate of soda to take up a portion of the acid of the alum. Acetate of lime is frequently substituted for acetate of lead, in the preparation of "buff-liquor."

The buff colour in the specimen following (No. 14) is produced by iron buff.

To impart a buff ground, the pieces are padded in

No. 14.

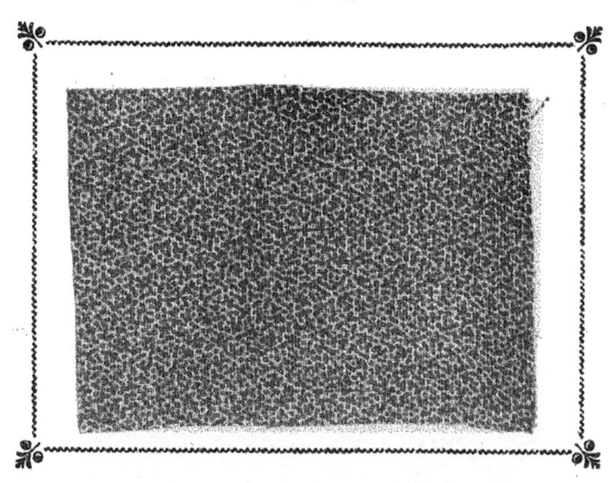

buff-liquor of any strength between 2° and 10° Tw., according to the shade of colour desired, then dried by being drawn either through the hot-flue or over iron boxes filled with steam, and aged for one or two days. Some printers then wince the pieces in water containing some chalk, and afterwards pass them through a solution of carbonate of soda; but it is much better to pass them at once through a solution of caustic soda, or through milk of lime.

During the ageing of the padded goods the salts of the protoxide of iron become subsalts of the peroxide, which are decomposed in the alkaline or calcareous solution, the acids being withdrawn by the alkali while the peroxide of iron becomes fixed on the cloth.

To obtain an iron buff figure, the pieces may be printed with a buff-liquor of any strength between 10° and 30° Tw. thickened with either gum, calcined farina, salep, or British gum. After being dried and aged, the pieces are passed directly through a solution of caustic soda, or else through milk of lime.

Manganese bronze.—The brown ground of the specimen annexed (No. 15) is produced by manganese bronze or peroxide of manganese.

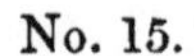

A solution of manganese sufficiently pure for producing the bronze, may be obtained from the residue of the process for chlorine, by saturating the remaining free sulphuric or muriatic acid with chalk, allowing the precipitate to settle, and decanting and concentrating the clear supernatant liquid. The chalk serves not only to saturate the free acid, but to precipitate peroxide of iron from the soluble salts of that oxide which this bye-product always contains. Lime has been recommended for this purpose instead of chalk, but it is never employed on the large scale, as an excess would decompose the salts of manganese as well as those of iron; an excess of chalk, however, is without action on the manganese salt.

A purer solution of manganese is prepared by heating the residue of the chlorine process with more black oxide of manganese until the evolution of chlorine almost ceases, and then adding, either chalk, or freshly precipitated carbonate of manganese, until the liquid becomes colourless. Having been allowed to settle, the solution is decanted and concentrated by evaporation.

To impart a dark bronze ground such as that in the above specimen, the strength of the solution of chloride of manganese may be about 26° Twad. For lighter shades, the solution may be made as weak as 4° Twad.

After having been padded and dried, the goods are passed through a cold caustic ley, whereby protoxide of manganese becomes precipitated on the cloth. On exposure to the air, the protoxide soon absorbs oxygen, passing into the state of the brown peroxide; but the peroxide may be produced immediately by wincing the goods in a solution of chloride of lime or chloride of soda as soon as they are taken out of the caustic ley. The common practice is to expose the pieces to the air until they acquire a good full colour, and then to complete the peroxidation of the manganese by a dilute solution of chloride of lime.

Peroxide of manganese is very seldom applied as a figure on a white ground. The solution of the chloride used for this purpose may have a density about 16° Tw., and be thickened with from two pounds to two pounds and a half of gum to the gallon. A small quantity of tartaric acid is a useful addition to such a solution. The printed and dried cloth is drawn through a caustic ley, exposed to the air, and winced in a solution of chloride of lime as above.

Scheele's green.—To obtain a ground of Scheele's green (arsenite of copper), the cloth is padded two or three times with a solution of nitrate of copper, or else with a mixture of the sulphate and acetate containing a little size, and after being dried, is winced in a dilute solution of a caustic alkali, to fix the oxide of copper. The cloth is then rinsed in water and winced in a dilute solution of arsenious acid, or else in a solution of arsenite of soda.

For the manner of applying a few other mineral colours, namely, antimony orange, arseniate of chromium, orpiment, and prussiate of copper, see page 254, *et seq.*

Mineral colours are frequently combined with steam and madder colours in the same design. When this is the case, the madder colours are always applied first, the mineral colours next, and the steam colours last.

The following method of procuring a design in black, purple, two shades of red, two shades of buff, green and yellow, on a white ground, is an example of the combinations of mineral colours with madder and steam colours:

1. Print the cloth by the four-colour machine with the mordants for black, purple, and two reds (see pages 353 and 354);

2. Age, dung, dye in the madder-beck, clear and dry;

3. Print by the two-colour machine (or else by

blocks, according to the design) with buff-liquor of two strengths, thickened with starch or British gum;

4. Wince the cloth, after being aged, in milk of lime, to raise the buff, and rinse in water;

5. Dry and print by blocks with the mixtures for steam blue (page 374), and steam yellow (page 372);

6. Age, steam, and rinse.

A pleasing pattern may be obtained by combining in one design, on a white ground, figures or bars in different shades of iron buff, with a figure or stripe in steam blue. The buffs are first applied in the usual manner.

4. RESIST STYLE.

The object of the resist style of work is to produce a white or coloured design on a coloured ground by the topical application, in the first place, of a substance called the *resist*, *reserve*, or *resist paste*, which has the property of preventing the attachment or developement of colour, when the whole surface of the cloth is afterwards impregnated with a dyeing material. One class of resists, consisting of substances of an unctuous nature, acts merely mechanically; another class acts both mechanically and chemically. The latter kind are divisible into two subdivisions, according as their influence is exerted merely on the mordant or on the colouring matter itself.

1. *Fat resists.*—Resists of an unctuous nature are chiefly used for applying to goods of silk and wool,

but they may be also advantageously applied, in particular circumstances, to goods of cotton. The annexed specimen (No. 16) is an example of the

No. 16.

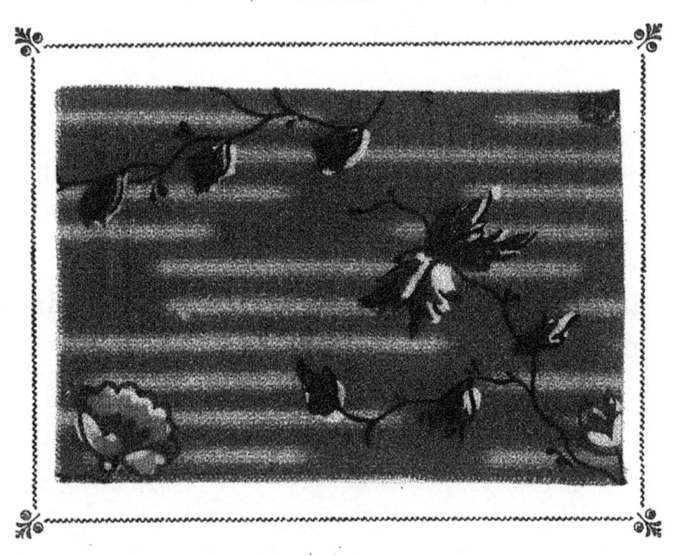

combination of such a style of work with madder colours and steam colours. The specimen No. 7, page 355, shews the appearance of the above in an early stage of the process. After having been printed, dyed, and cleared as already described, the red and lilac figures are covered (or overlaid) with a resist consisting, usually, of an intimate mixture of suet and gum-water.* The whole is then run over by the roller with weak iron liquor for the lilac ground, and the cloth is aged, dunged, dyed, and cleared. The mixtures for steam green and steam yellow are afterwards pegged in by blocks, and the steaming is performed in the usual manner. The

* A solution of citrate of soda (obtained by neutralizing lime-juice with soda), thickened with pipe-clay and gum, might be used instead of the mixture mentioned in the text.

mixture for the green in the above specimen is quite similar to that described at page 379; that for the yellow was made by mixing a quart of red liquor of spec. grav. 8½° Tw. with a gallon of berry liquor of 2° Tw.

In this style of work, however, the dyeing with madder might as well be performed at one operation, as the red and lilac mordants for the figures are not at all injured by the fat resist with which they are covered.

2. *Resists for mordants.*—The material generally used for preventing the deposition of a mordant on particular parts of the cloth is an acid or acidulous salt capable of uniting with the base of the mordant, to form a compound soluble in water and not decomposable into an insoluble subsalt during the hanging of the mordanted goods, previous to dunging and dyeing. The resist commonly employed for the iron and aluminous mordants is lemon-juice or lime-juice, or a mixture of one of these with tartaric and oxalic acids and bisulphate of potash. The thickening material is either a mixture of pipe-clay or china-clay with common gum, a mixture of British gum with gum senegal, or British gum alone. Lemon-juice or lime-juice is decidedly preferred to pure citric acid (which is the acid principle of these juices), as the mucilaginous matters in the former impede the crystallization of the acid within the pores of the cloth, and thus render it better adapted to prevent the attachment of the mordant in an insoluble form. The strength of the resist is regulated by the strength of the mordant afterwards applied it is seldom used of a higher density than 2° Twad.

The specimen (No. 17), exhibiting a design in black, lilac and white on a lilac ground, was produced

No. 17.

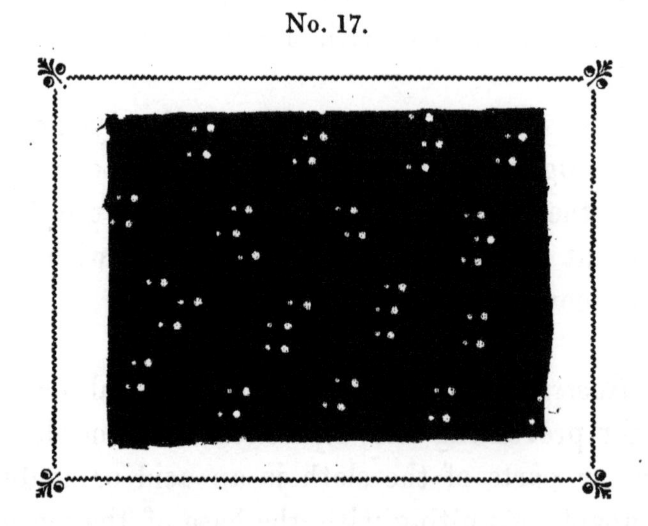

by adapting the resist style of work to madder colours. The printing for such a pattern may be performed by the three-colour machine in the following order:

By the first roller; the resist, which may be either lemon-juice of spec. grav. 2° or 3° Twad., thickened with four pounds of British gum to the gallon, or else a solution of about the same density, of tartaric and oxalic acids in weaker lemon-juice, also thickened with British gum:

By the second roller; the mordant for the black, which is iron liquor of spec. grav. 8° Twad., thickened with a pound and a half of flour to the gallon:

By the third roller; the mordant for the ground of lilac, which is iron liquor of spec. grav. $1\frac{1}{2}$° Twad., thickened with four pounds of British gum to the gallon.

The application of the mordant for the ground may be made by the padding machine (fig. 41, page

384), but it is commonly done by the cylinder machine, the entire surface of the copper roller being slightly roughened or engraved in close diagonal lines, so as to enable it to afford an uniform deposit on the cloth.

The operations of ageing, dunging, dyeing, and clearing are conducted in much the same manner as if the acid resist had not been applied. It is usual, in this style of work, to add a small quantity of chalk to the dung-beck, in order to counteract the effects of the free acid in the resist.

Iron liquor may be resisted or prevented from affording a deposit of insoluble subsulphate during the ageing, by a process somewhat different from the above, the resisting agent being protochloride of tin (commonly called salts of tin), instead of a free acid or an acidulous salt. A mixture of protochloride of tin and iron liquor does not afford a deposit of a subsalt of iron during the ageing of cottons printed with the mixture, probably through the occurrence of a double decomposition with formation of acetate of tin and chloride of iron. The latter compound does not afford an insoluble precipitate during the ageing, and may be entirely removed from the cloth by washing.

When a piece of cotton cloth is printed with a solution of salts of tin by the first roller of a two-colour machine, and with iron liquor by the second roller, over the parts printed by the first roller, such a mixture as the above is of course formed wherever the salt of tin had been applied, and no subacetate of iron is deposited there during the ageing.

The protochloride of tin, however, is never applied in this way with a view of producing a white figure on a coloured ground; it is commonly mixed with red liquor, as the deposition of the insoluble sub-sulphate of alumina from that preparation is not interfered with by the protochloride. After a piece of cloth thus printed has been aged, dunged, dyed in the madder-beck, and cleared, it therefore presents a red figure surrounded by purple or lilac. It is to be observed that this method of procedure is only followed when a better definition of the red design is required than could be attained by leaving a blank figure in the roller for the iron liquor, and afterwards printing the red liquor on the white parts either by a second roller or by the block. To resist weak iron liquor and impart the mordant for a full red with madder, the mixture may have the following composition:

1 gallon of red liquor of 18° Twad.,
4 oz. crystals of protochloride of tin; with a sufficient quantity of British gum or a mixture of British gum and starch as the thickener.

To obtain a design in full red and black on such a lilac ground as the specimen No. 16, page 394, contains, the cloth is printed with strong iron liquor for the black, with the above mixture for the red, and with iron liquor of 1° Twad. for the lilac; after which it is aged, dunged, dyed, &c. in the usual manner. A great variety of pleasing effects may be produced by combining this kind of work with steam or topical colours, the iron liquor not being applied as a ground, but as a design extending on each side of the red figure, and on the parts left white the steam colours are applied, after dyeing with madder and clearing.

Another material, much used as a resist for red liquor and iron liquor, is a solution of citrate of soda, prepared by neutralizing lime-juice of about 4° Twad. with soda, thickened with a mixture of gum and pipe-clay. The action of this resist may probably be referred to the tendency of citric acid, like oxalic acid and a few others, to form a double salt with peroxide of iron or alumina and an alkali, which affords no precipitate of alumina or oxide of iron during the ageing. In this case a portion of the alkali in the neutral citrate is withdrawn by the acetic acid in the mordant, an acid citrate of soda being thus formed. Neutralized lime-juice of 4° Tw. has about the same resisting power as the unneutralized juice of 2° Tw.

The principal use of neutralized lime-juice as a resist for iron liquor is to protect figures previously applied in madder colours; for which purpose the free acid is quite inapplicable, as it would dissolve the mordant on the cloth in combination with the colouring matter. If the dyeing of the cloth with madder for such a pattern as that of the specimen No. 16, page 394, is performed at two operations, neutralized lime-juice is generally preferred to the fat resist for protecting the red, lilac and white figures of the specimen No. 7, page 355; but if the maddering is to be completed at one operation, the fat resist must be used.

3. *Resists for the colouring matter.*—The production of a white or coloured pattern on a coloured

ground by the direct action of a resist on a colouring matter, is chiefly practised with indigo, at least in the printing of calicos. The substances most commonly employed for this purpose are salts of the black oxide of copper, particularly the sulphate and the acetate. The other substances employed as resists for indigo are sulphate of zinc, chloride of zinc, chloride of mercury (corrosive sublimate), and a mixture of corrosive sublimate and binarseniate of potash. None of these are in common use except sulphate of copper, acetate of copper, and sulphate of zinc.

The ordinary course of operations practised in this style of work, with the view of producing merely a white pattern, are the following:

The resist, mixed with unctuous matters and properly thickened, is first printed on such parts of the cloth as are not to absorb the indigo, and the goods are suspended for one or two days (according to the composition of the resist) in a chamber at common temperatures, and not very dry. The pieces are then fixed on a frame and dipped into the indigo vat, which contains, in solution, the colourless hydruret of indigo, or indigotin (see page 244). The solution of indigotin is immediately absorbed by the cloth on all parts where the resist had not been printed, which parts become deep blue when the cloth is afterwards exposed to the air, the soluble indigotin passing into the state of insoluble indigo-blue through the absorption of oxygen. But wherever the resist had been applied, the solution of indigotin is not absorbed by the cloth, partly on account of the unctuous matters contained in the resist, but chiefly from the action of the metallic salt on the indigotin in solution, by which either indigotin or else indigo-blue becomes precipitated before the solution reaches the interior of the fibre, and being precipitated, the indigo is rendered incapable of being absorbed by the pores of the cloth. The indigo-blue which is formed by the resist is merely attached to the surface of the cloth, and is easily removed by washing.

The first chemical change which occurs when the cloth printed with a resist containing sulphate or acetate of copper or corrosive sublimate is dipped into the indigo-vat, is the decomposition of the metallic salt in the resist by the alkali or lime in the vat, whereby the cupreous salts afford a precipitate of hydrated protoxide of copper (black oxide) and corrosive sublimate, a precipitate of protoxide of mercury (red oxide).* These oxides are no sooner produced than they exert an oxidizing action on the indigotin in solution, and become reduced to the state of inferior oxides. The protoxide of copper becomes the red or suboxide (which generally requires to be cleared away at the end of the process by wincing the goods in dilute sulphuric acid), and the protoxide of mercury becomes the black oxide or suboxide.

The mode of action of sulphate of zinc is somewhat different. This salt exerts no oxidizing action on the indigotin, but causes the precipitation of that substance in an insoluble state by withdrawing the lime which holds it in solution. By the reaction of the lime in the solution on the sulphate of zinc, there are also precipitated sulphate of lime and oxide of

* Commonly called the peroxide.

zinc, both of which offer a mechanical impediment to the access of the liquid to the cloth.

The sulphate of zinc resist should be used only in cases where it would not be allowable to expose the goods to dilute sulphuric acid, after having been dipped into the blue-vat, as if, for instance, a design had been previously applied in madder colours, this operation being unnecessary with the zinc salt. Corrosive sublimate is far too expensive, and also too weak, to be commonly employed as a resist for indigo, except where a very delicate figure in madder colours has to be protected.

The white pattern in the annexed specimen (No. 18) was produced by such a process as that just

No. 18.

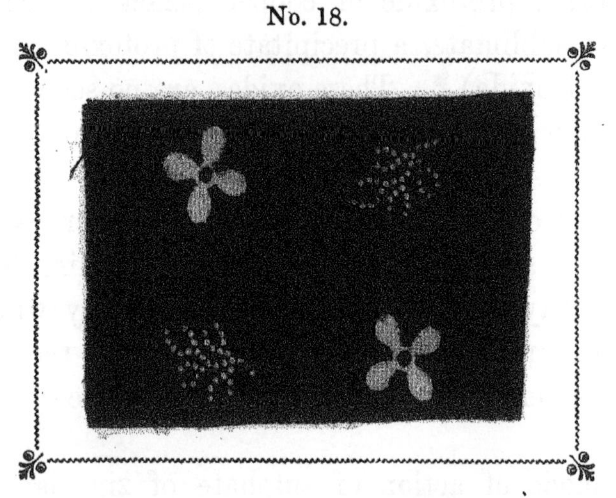

described. For this style of work, the blue-vat or solution of indigotin may be made by mixing one hundred pounds of ground indigo, one hundred and thirty-five pounds of copperas, one hundred and seventy-five pounds of lime, and from sixteen hundred to two thousand gallons of water. The vat is fit for use two days after the materials are mixed.

For a deep blue, the cloth is dipped into the vat for ten minutes and then exposed to the air for the same length of time, and the dipping and exposure to the air are repeated until the required depth of colour is obtained.

The composition of the resist paste is varied according to the depth of colour in the blue ground. The following mixture is well adapted for dark blue:

No. 1.

3 to 4 pounds of sulphate of copper,
7 pints of water,
5 pounds of pipe-clay, china-clay, or sulphate of lead,
4 ounces of soft soap,
3 pounds of gum.

For a resist paste for light blue, the proportion of sulphate of copper may be reduced to eight ounces in a gallon of the paste. This resist we may call No. 2.

The sulphate of zinc resist, for protecting a design in madder colours as well as for preserving some white, may have the following composition:

No. 3.

4 to 5 pounds of sulphate of zinc,
2 quarts of boiling water,
5½ pounds of pipe-clay,
4 ounces of soft soap,
2 ounces of hog's lard,
2 quarts of gum senegal water, containing six pounds of gum to a gallon of water.

The sulphate of zinc is first dissolved in the hot water, and with this solution, while warm, the pipe-clay, soap, and lard are thoroughly incorporated. When the mixture is cold the gum-water is added.

Such are the methods of obtaining a white figure on a blue ground by the resist style. To procure a design in white and light blue on a dark blue ground, the cloth is first printed with the strong cupreous resist (No. 1), dipped in the blue-vat and cleaned, as if a white design only is required. After being dried, it is printed with the weaker resist containing sulphate of copper (No. 2), again dipped in the blue-vat to a lighter shade, cleared in dilute sulphuric acid and dried.

A great variety of coloured designs on the same ground may also be obtained by combining with the resist, either one of the saline solutions capable of imparting a mineral colour, or else the mordant for a colouring matter to be applied by the madder style.

A design composed of yellow figures on an indigo ground, is very commonly and easily obtained by combining the resist with a salt of lead, and padding or wincing the cloth in a solution of bichromate of potash after being dipped into the indigo-vat and cleared. The successive operations to which a piece of calico is subjected in this kind of work are the following:

1. Printing with the mixture of resist and salt of lead, which may have the following composition:

> 1 gallon of water,
> 3 to 4 pounds of sulphate of copper,
> 1 pound of nitrate of lead,
> 1 pound of acetate of lead,
> 3 pints of a paste of precipitated sulphate of lead,
> 5 or 6 pounds of pipe-clay,
> 2 to 3 pounds of gum.

2. Hanging for one or two days in a room having a rather humid atmosphere;

3. Dipping into the indigo-vat;

4. Passing through dilute sulphuric acid;

5. Steeping in water for half an hour, and washing;

6. Wincing in a dilute solution of carbonate of soda;

7. Wincing in a solution of bichromate of potash, containing five ounces of the bichromate per piece of calico;

8. Wincing in dilute muriatic acid;

9. Washing in water.

To obtain a figure of chrome-orange instead of chrome-yellow, the calico may be first treated as above, and afterwards winced in hot milk of lime to convert the chrome-yellow into chrome-orange.

To procure a design in yellow and light blue on a dark blue ground, the cloth is submitted to the following operations:

1. It is first printed with the mixture of sulphate of copper and salts of lead for chrome-yellow, and on the parts to be light blue with a mixture of sulphate of copper and acetate of copper, formed by mixing solutions of acetate of lead and sulphate of copper, allowing the mixture to settle and decanting the supernatant liquid;

2. After being dried, the cloth is dipped into the blue-vat for the dark ground;

3. It is next passed through dilute sulphuric acid to clear the whites of the suboxide of copper, and washed in water;

4. After being winced in a mixed solution of carbonate of soda and carbonate of ammonia, it is dipped a second time into the blue-vat for the light blue of the figure, and then washed in water;

5. It is afterwards winced in a solution of bichromate of potash, and then drawn through a cistern containing a solution of one ounce of oxalic acid and one ounce of sulphuric acid to the gallon of water.

A pattern comprising a figure of iron buff on a ground of indigo may be applied to cloth by a similar combination of the padding and resist styles, the cupreous resist being mixed with a salt of the peroxide of iron. After the ground of indigo is applied, the cloth is slightly washed, and then winced in a warm dilute solution of carbonate of soda to precipitate hydrated oxide of iron. A buff figure on a dark green ground is sometimes produced by first printing the cloth with the white resist No. 1 (page 403), then dipping into the blue-vat, and after the cloth is cleared and dried, padding it with buff-liquor and raising the buff by carbonate of soda.

Another method of producing a coloured figure on the indigo ground is by combining with the resist paste a mordant for a vegetable colouring matter, to be applied by the madder style, after the cloth has been dipped into the indigo-vat. This kind of work, which is susceptible of a great variety of modifications, is distinguished as the *lapis* or *lazulite style*, from the resemblance of the calico thus printed and dyed to the mineral lapis lazuli. It is also known as the *neutral style.*

To obtain a red figure on the indigo ground, the cloth is printed with a resist paste composed, essentially, of red liquor, sulphate of zinc, and acetate of copper.

This resist may be made of the following materials, mixed in the order in which they are placed:

No. 1.

2 gallons of boiling water,
6 pounds of alum,
4 pounds of crude acetate of lead,
4 ounces of chalk, added in small quantities at a time, and
6 ounces of sulphate of zinc.

These materials having been thoroughly incorporated, the mixture is allowed to settle, and the clear supernatant liquid is decanted and mixed with acetate of copper and gum senegal, thus:

No. 2.

1 gallon of the above clear liquid,
3 ounces of acetate of copper,
18 ounces of gum senegal,
5 pounds of pipe-clay,
4 ounces of soft soap, and a little ground indigo for "sightening."

One half of the liquid is well mixed with the acetate of copper, pipe-clay, and soap, and the gum senegal is afterwards added, dissolved in the other half of the liquid.

After being printed with this resist, the cloth is aged for two or three days, and then subjected to the following operations:

1. Drawing by rollers once* through the blue-vat at 70° Fahr., made from the proportions of lime, copperas, and indigo mentioned at page 402, but with only eight hundred gallons of water;
2. Rinsing in water;
3. Dunging or branning;
4. Washing at the dash-wheel;

* If the cloth is exposed to the blue-vat for some length of time, the aluminous mordant is separated from the cloth by the lime in the vat.

5. Dyeing in the madder-beck, with from two to five pounds of madder per piece;

6. Clearing by boiling first in bran-water and afterwards in soap-water.

To produce a light red figure with madder, the resist may have the following composition:

4 measures of the sulphate of zinc resist paste No. 3, page 403,
1 measure of the mixture of red liquor and sulphate of zinc made as above,
1 measure of weak peachwood liquor,
1 measure of water.

For two reds the cloth may be printed with the preceding mixtures at the same time by the two-colour machine, and be treated afterwards in the manner just described.

To obtain merely a small black figure on the indigo ground the cloth may be dipped in the blue-vat to the required shade, and then be printed with the mixture for producing a topical black dye, such as pernitrate of iron, copperas, and extract of logwood (see page 360). But if the design includes figures in red and white, the black forming a more considerable portion of the figure than a mere outline, it is better to mix iron liquor of 8° or 10° Twad. with the cupreous resist and to dye the cloth in the madder-beck, after having dipped it into the blue-vat to the proper shade. A great variety of purple, lilac, and chocolate tints may also be obtained on the same ground, by combining with the cupreous resist, either weak iron liquor or else mixtures in various proportions of iron liquor with red liquor, and dyeing in madder after the dipping in the blue-vat.

To impart to the blue ground a design in light blue, together with a colour capable of being applied by the madder style, the cloth may be treated as follows:

1. Print with the white resist No. 1 (page 403);

2. Dip into the blue-vat, wash, wince in dilute sulphuric acid, rinse in water and dry;

3. Print the mixture of the mordant with the cupreous resist on a part of the white figure produced by the first resist;

4. Dip a second time into the blue-vat, to obtain a light blue on the parts not protected by the second resist, rinse in water;

5. Dung, wash, and dye in the decoction of the dye-stuff, and afterwards clear by branning.

If a white figure is required in addition to the above, the cloth is first printed with the strong white resist and dipped into the blue-vat as already described, and is afterwards printed on the protected parts, by the two-colour machine, if the design admits, with the mixture of mordant and salt of copper, and also with a mild resist such as No. 2, page 403. It is then dipped into the blue-vat and dyed in the usual manner.

To procure such a pattern as the annexed (No. 19), containing a design in orange, crimson, and white on a blue ground, the cloth is printed by the two-colour machine with the mixture of salts of copper and salts of lead (page 404) on the parts to be orange, and with a white resist on the parts to be crimson and white. After being dipped into the blue-vat and cleared in dilute sulphuric acid, it is winced in the following liquids: 1°, solution of car-

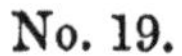

No. 19.

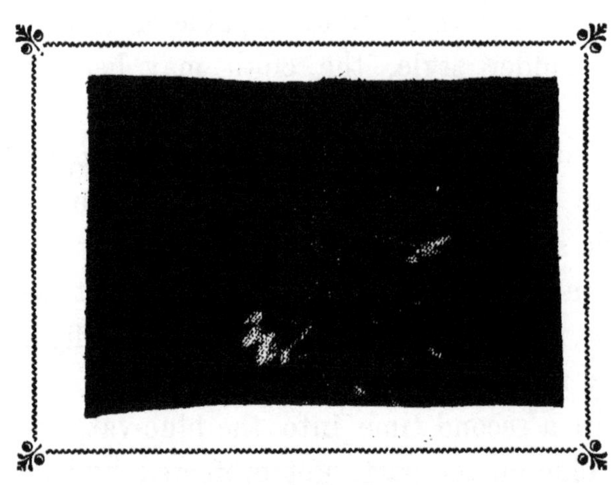

bonate of soda; 2°, solution of bichromate of potash; and 3°, dilute muriatic acid. It is next passed through hot milk of lime to convert the chrome-yellow into chrome-orange, rinsed and dried, and is afterwards printed by block on parts of the white with the mixture for a topical or steam red or crimson.

The following specimen (No. 20), exhibiting a design in blue, yellow, green, red, and white on a dark chocolate ground, was produced by combining the lazulite style with a topical colour. This kind of work is distinguished as the "chocolate ground neutral style." For such a pattern the cloth is first printed (either by machine or block) with the white resist,* No. 1, page 403, on all the parts required to be yellow and white; with the mixture of red liquor,

* If a very small or well-defined white figure is required, the cupreous resist should be mixed with lime-juice and sulphuric acid or bisulphate of potash, to resist the mordant in the chocolate resist, afterwards applied as a blotch. Such a mixture is designated (not very appropriately) *neutral paste.*

No. 20.

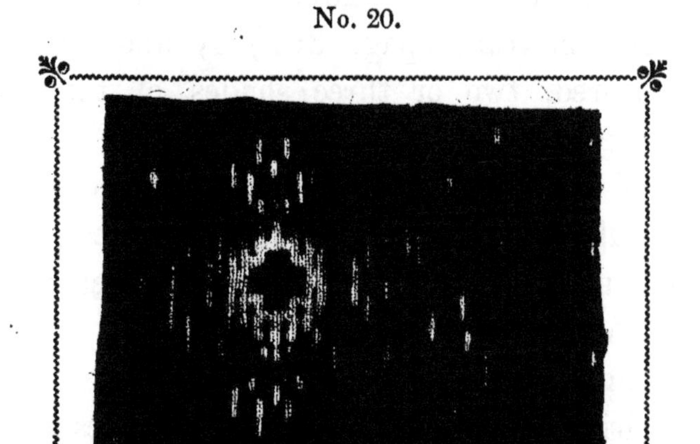

sulphate of zinc, and acetate of copper, made as described at page 407, on the parts required to be red; and with a mixture of iron liquor, red liquor, sulphate of copper, and soft soap thickened with pipe-clay and gum, for the chocolate ground or "blotch." After having been aged for a day or two the cloth is drawn once through the indigo-vat, then washed, dunged, dyed in the madder-beck, and cleared by branning. Lastly, the mixture for a topical or steam yellow is applied by the block.

A process referable to the resist style is that by which a white figure is obtained on a ground of catechu brown. On the parts to be preserved white, the cloth is printed with a solution of citrate of soda (such as that obtained by exactly neutralizing lime-juice with caustic soda) thickened with a mixture of pipe-clay and gum; or else, what is preferred, a mixture of sulphate of zinc, pipe-clay, and gum. Such a resist may be printed on the cloth by one roller of a two or three colour machine, and the catechu mixture (page 361) by another roller, or if required, two or three shades of the brown may be applied by as many rollers. The action of both of these resist pastes is chiefly mechanical; but the sulphate of zinc also acts by precipitating the catechu in solution, and thus preventing its access to the fibre of the cloth.

The same resist may be employed for preventing the deposition of catechu on a coloured design previously applied in madder colours. The specimen No. 8, page 361, is an example of such a combination of the topical, madder, and resist styles. To produce this pattern, the cloth is treated as follows:

1. Print by the two-colour machine with strong red liquor for the red, and with a mixture of iron liquor with a little red liquor for the dark puce;
2. Age, dung, and wash in the usual manner;
3. Dye in the madder-beck;
4. Clear by branning, &c., and dry;
5. Cover all the figures thus produced, with a resist paste of sulphate of zinc, pipe-clay, and gum;
6. Apply the catechu ground by a roughened roller, and age for a couple of days, previous to washing at the dash-wheel.

A red shade may be given to the catechu ground by impregnating the entire surface of the cloth with weak red liquor, by a roughened roller, at the same time as the strong red and puce mordants are applied. Two shades of brown are sometimes imparted by applying the weak red liquor to certain parts of the ground, as for instance in broad stripes, the intervening spaces having catechu only.

5. DISCHARGE STYLE.

The manner of producing a white or coloured pattern on a coloured ground by the topical application of a "discharger" to a cloth already mordanted or dyed is applicable to both mineral and vegetable colouring matters. Like the resist paste in the preceding style, the discharger may act either on the colouring matter itself or on the mordant before the cloth is exposed to a dyeing liquid. Dischargers for mordants are generally acid mixtures quite similar to resists for mordants, but dischargers for colouring materials are obtained from different classes of chemical substances according to the nature of the colouring matter to be removed. The essential property required in a discharger is that of converting the substances on the cloth into colourless or soluble products, which may be removed from the cloth so as not to interfere with the subsequent application of a colouring material to the parts discharged.

1. *Dischargers for colouring matters.*—The materials used as dischargers for vegetable colouring principles are chlorine and chromic acid, the bleaching powers of which have before been alluded to (pages 238 and 239).

To effect the topical discharge of a vegetable colouring matter by means of chlorine, with the production of a white figure, the dyed cloth is printed on those parts which are to be discharged, with a thickened acid mixture, the composition of which is varied according to the fastness of the colour to be destroyed, and after being suspended to dry for a day or two, the cloth is drawn (by a pair of squeezing rollers) through a solution of chloride of lime not stronger than 8° Twaddell or 1·040. The calico should be extended on rollers while being drawn through the solution, and should not occupy more than two or three minutes in its passage. The solution of chloride of lime is usually contained in a rectangular cistern of wood lined with lead, of the following dimensions; six or eight feet long, three feet wide, and four or five feet deep. As soon as the goods are taken out of the solution of chloride of lime, they are put to soak in water; after which they are washed either at the dash-wheel or in the rinsing machine, and then dried.

The chemical reactions which take place in this process are by no means complicated. Chloride of lime does not of itself bleach Turkey red and some other fast colours immediately; so that a cloth dyed with such colours may remain for some minutes in contact with a solution of chloride of lime without any deterioration in colour. But the acid applied to certain parts of the cloth combines with the base of the chloride and liberates free chlorine, which exerts an instantaneous bleaching action on the vegetable colouring matter on those parts of the cloth. Almost the only colours to which chlorine can be thus applied as a discharger, are Turkey red and other madder colours and indigo, as the more delicate colours are easily discharged by chloride of lime alone.

A white discharger adapted for all madder colours except Turkey red may be made by dissolving four pounds of tartaric acid in a gallon of water, mixing this solution with a gallon of lime-juice of spec. grav. 44° or 48° Twad., and thickening the mixture with pipe-clay and gum.

The white discharger for Turkey red requires to be somewhat stronger than the above. It may be made by mixing four pounds of tartaric acid with a gallon of lime-juice at about 30° Twad., and after thickening with pipe-clay and gum, adding about a pound of concentrated sulphuric acid, or two pounds of bisulphate of potash.

In a particular style of work, the Turkey red is discharged by the direct topical application of chlorine, or rather of an aqueous solution of chlorine. It is in this way that the celebrated Bandana handkerchiefs, which have white figures on a dark ground, have been most successfully imitated by Messrs. Monteith of Glasgow. The style is only practised in the manufacture of handkerchiefs.

From ten to fourteen pieces of cloth, previously dyed Turkey red, are stretched over each other quite parallel, and passed together by portions at a time (proceeding from one end of the pieces to the other end), between two leaden plates, one of which is superimposed immediately over the other. Each of these leaden plates is cut completely through so as to lèave hollow places on all the parts required in white on the red ground. By means of a hydraulic press, the pieces of cloth are compressed between the leaden plates with a force of three hundred and twenty tons on the whole surface. While the cloth is exposed to this immense pressure, an aqueous solution of chlorine (obtained by adding sulphuric acid to a solution of chloride of lime) is made to percolate downwards through the pieces by the openings in the leaden plates. As the compressed state of the cloth prevents the imbibition of the liquid except by the parts opposed to the design on the lead, the solution passes on in a circumscribed channel to the lower leaden plate, where it escapes and is conveyed away by a waste-pipe. The portions of cloth through which the liquid passes are entirely deprived of their colour.

As soon as the chlorine solution is passed through, water is made to percolate in a similar manner to wash away the chlorine, else the definition of the pattern would be impaired. The passage through the cloth of the chlorine solution and the water for washing is sometimes assisted by a pneumatic apparatus consisting of a large gasometer, from which a current of air is caused to proceed under a moderate pressure, and act in the direction of the liquid.

When a considerable quantity of water has passed through the cloths, the pressure is removed and the pieces are washed and slightly bleached, whereby the lustre both of the design and ground is considerably increased.*

After the production of a white figure on a coloured ground by the application of the acid discharger and immersion in the solution of chloride of lime, coloured figures may be applied either to

* A detailed account of this interesting process, by Dr. Ure, may be found in the Journal of the Royal Institution for 1823.

the ground or to the white figure by grounding in topical colours by the hand-block. A common method of imparting a coloured figure is by mixing with the acid discharger one of the two solutions necessary for producing a mineral colouring material. For example, to impart a yellow figure to a piece of cotton dyed with Turkey red, such as the specimen No. 21, the cloth is treated in the following manner:

No. 21.

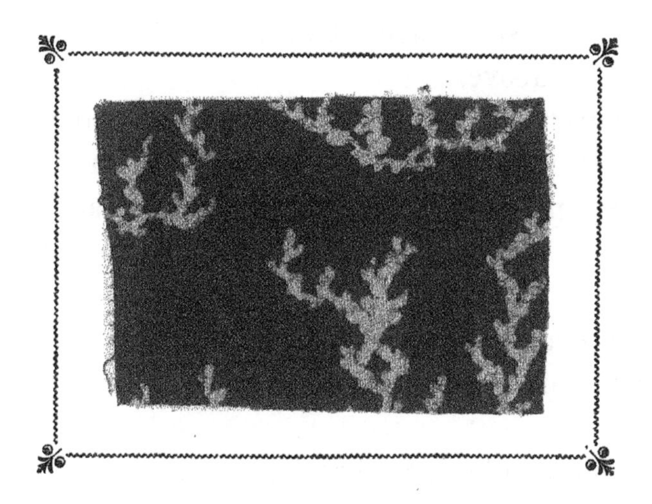

1. It is printed by the machine with a chrome-yellow discharger composed of

> 1 gallon of lime-juice of spec. grav. 20° Twad.,
> 5 pounds of tartaric acid,
> 4 pounds of nitrate of lead, with a mixture of Pipe-clay and gum as the thickener.

2. After hanging for a day or two, the piece is passed through the solution of chloride of lime at 8° Twad.

3. It is soaked in water and then slightly winced in water.

4. The piece is next winced for about a quarter of an hour in a solution of bichromate of potash containing from three to five pounds to the piece.

5. It is lastly passed through, or winced in, dilute muriatic acid, washed at the dash-wheel and dried.

To obtain both a white and a yellow figure on a Turkey red ground, the dyed cloth may be printed with two acid dischargers, one intended for the production of the white figure, the other for the yellow figure. The subsequent treatment of the cloth is the same as above.

To impart a blue figure to the same ground, the dyed cloth is printed with a mixture of soluble Prussian blue, permuriate of tin, and tartaric acid, after which it is drawn through a solution of chloride of lime. The Turkey red thereby becomes discharged and the Prussian blue fixed on all the parts where the above mixture had been printed.

The only substance besides chlorine which can be conveniently employed to effect the topical destruction or removal of a vegetable colouring matter is chromic acid, which produces the decomposition of the colouring matter by virtue of its oxidizing power, the chromic acid becoming reduced to the state of green oxide of chromium. The vegetable colouring principle best adapted to this kind of work is indigo.

To obtain a white pattern on an indigo ground by means of chromic acid, the cloth is first dyed uniformly with indigo in the ordinary manner, and then padded with a solution of bichromate of potash containing about five or six ounces per piece. After being carefully dried in the shade at the ordinary temperature, the cloth is next printed with a discharger containing tartaric acid, oxalic acid, citric acid, and sometimes muriatic acid; and immediately after the impression it is winced in water containing some chalk in suspension, then washed at the dash-wheel, passed through dilute sulphuric acid, and lastly washed in clean water.

The colour of the indigo on the cloth is destroyed immediately on the application of the acid discharger: chromic acid is then liberated from the bichromate through the superior affinity of the acids in the paste for the potash, and the free chromic acid at once oxidizes and destroys the colouring matter. Indigo is almost the only substance which can be adapted to the chromic acid discharger, owing to the oxidizing action which the bichromate of itself exerts on vegetable colouring materials in general; hence the reason also for drying the dyed goods, after being padded with the bichromate, in a darkened chamber and at the ordinary temperature.

To produce a yellow instead of a white figure, the acid discharger may be mixed with a salt of lead: in other respects the process is the same as above.

The following method of obtaining a white figure on a dark green ground is an example of the combination of the madder style of work with the chromic acid discharge style.

1. Dip the cloth in the blue-vat to the desired shade;

2. Pad in a mixture of red liquor with bichromate of potash containing five or six ounces of the latter to the gallon, and dry in the shade;

3. Print the cloth, without being washed, with a mixture of lime-juice, sulphuric acid, and oxalic acid;

4. Pass the cloth through a mixture of hot water and chalk, and dye in a decoction of quercitron bark;

5. Wash and clear by branning.

In this process, the mixture of lime-juice, sulphuric acid, and oxalic acid, not only liberates chromic acid from the bichromate of potash, but also dissolves the subsulphate of alumina deposited from the red liquor: the parts on which this mixture is applied do not therefore become permanently dyed yellow when the cloth is exposed to the decoction of quercitron.

The discharge style is applicable to cloths dyed with mineral as well as with vegetable and animal colouring matters.

1. A white figure may be produced on a ground of Prussian blue, by imprinting on the cloth a paste containing a caustic alkali (either potash or soda), and passing the cloth afterwards through a solution of oxalic acid. The Prussian blue is here decomposed by the action of the alkali, affording yellow prussiate of potash, or prussiate of soda, which may be removed by washing, and peroxide of iron, which is precipitated on the cloth, but is afterwards dissolved out by the oxalic acid.

2. A white figure on a ground of manganese brown may be very readily obtained by imprinting the cloth, after being dyed brown in the ordinary manner (see page 390), with a slightly acid solution of protochloride of tin of a specific gravity about 70° or 80° Twad., or containing a pound and a half or two pounds of the protochloride per gallon, according to the intensity of the shade of the manganese ground. The solution of protochloride of tin is thickened with about a pound of starch to the gallon. The peroxide of manganese on the cloth is decomposed by the protochloride of tin and converted into protochloride of manganese, which being a very soluble salt is easily dissolved out by washing, leaving the parts white, or nearly so, on which the salt of tin had been applied. Peroxide of tin is formed at the same time, and remains for the most part attached to the cloth, but being white, it does not vitiate the pattern, and if required, may be made subservient to the application of the colouring principle of a vegetable dye-stuff, as peachwood, quercitron, or logwood. As most acidulous mordants are capable of removing the peroxide of manganese and inserting their own bases instead, a great variety of coloured designs may be applied to the manganese ground by afterwards dyeing such goods in various dye-becks.

To impart a design in white, blue, and yellow on the bronze ground (such as the specimen No. 15, page 390), the cloth on which the manganese has been raised may be printed with the salts of tin for the white; with a mixture of berry liquor, alum, and salts of tin for the yellow; and with a mixture of salts of tin, prussiate of potash, pernitrate of iron, muriatic acid and British gum, for the blue spots. The colour of the latter mixture is at first greenish-white, but it changes to blue on exposure to the air.

A design in different shades of red and pink may be communicated to the same ground by means of a mixture of peachwood liquor or cochineal liquor with alum, perchloride of tin, and protochloride of tin, thickened with gum tragacanth; and a mixture of logwood liquor, with alum and the two chlorides of tin, thickened with starch, may be used for imparting different shades of purple and violet to the same ground.

A figure in chrome-yellow may be produced on a ground of manganese bronze by printing on the dyed cloth a discharging material composed of tartaric acid, nitrate of lead, and salts of tin. After the cloth is dried, it is passed, first, through lime-water, then through a solution of bichromate of potash, and afterwards through dilute muriatic acid to brighten the yellow.

3. Protochloride of tin, when mixed with sulphuric and tartaric or oxalic acid, is also used as the discharging material for chrome-yellow and chrome-orange. The discharge of the chromates of lead is effected, in this case, by the reduction of the chromic acid to the state of green oxide of chromium, which forms soluble salts with the acids.

A variety of coloured designs may also be applied by combining with the discharger, the materials for the production of a topical colour. Thus, a blue figure is sometimes produced by printing on the orange or yellow cloth a mixture of the two chlorides of tin, Prussian blue, and muriatic acid; a violet figure, by logwood liquor mixed with alum, tartaric acid, protochloride of tin and starch; and a red or pink figure, by a similar mixture containing peachwood liquor instead of logwood liquor.

Another material which may be used for discharging chrome-yellow and chrome-orange, with a view of producing a white figure, is a strong caustic alkaline solution, but protochloride of tin will generally be found more convenient and more effective.

4. A white figure on a ground of iron buff (page 388) is obtained by applying to the coloured cloth a mixture of tartaric and oxalic acids with lime-juice, thickened with pipe-clay or China-clay and gum. The acids dissolve the peroxide of iron, and the figure is obtained perfectly white by washing. The readiest way of discharging the iron is to apply the acid mixture after the cloth has been padded in the iron liquor, and before it is exposed to the alkaline solution to precipitate the peroxide. A solution of protochloride of tin in a dilute acid, thickened with starch, is also sometimes used as a white discharger for iron buff; and for producing coloured designs, the protochloride may be mixed with perchloride of tin and either logwood liquor, peachwood liquor, or berry liquor.

The following method of producing white and buff-coloured figures on a dark green ground is an example of the combination of such a process as the above with the resist style.

1. The cloth is printed with the white resist for the indigo-vat (No. 2, page 403);

2. It is dipped into the blue-vat, rinsed, and dried;

3. It is padded with rather weak iron liquor and aged;

4. A solution of tartaric and oxalic acids in lime-juice, thickened with pipe-clay and gum, is applied by the block to parts of the buff spots;

5. The cloth is washed in water holding chalk in suspension to remove the acid paste;

6. It is lastly winced in an alkaline solution, to raise the buff, and then washed.

The white figure is here produced by the discharge of the salt of iron from parts of the spots on which the indigo had been resisted; the buff figure is the remainder of those spots, and the dark green ground results from the mixture of the indigo with the buff.

2. *Dischargers for mordants.*—Another method of producing white or coloured figures on a coloured ground, referable to the discharge style of work, is by the removal of the mordant previous to the application of the colouring material. This method is particularly adapted to grounds of madder and logwood with an iron or aluminous mordant. The material used for the discharge of the mordant is usually a mixture of tartaric acid, oxalic acid, and lime-juice, the proportions of the constituents being varied according to the strength of the mordant to be discharged. The following mixture may be

used for discharging the mordant from a piece of cloth impregnated with red liquor of spec. grav. 7° Twad. or weaker, or with iron liquor of spec. grav. 2° Twad. or weaker:

1 gallon of lime-juice of spec. grav. 6° Twad.,
3½ ounces of oxalic acid, and
4 ounces of tartaric acid,
Thickened with pipe-clay and gum if for application by the block, or with British gum if by the roller.

Sometimes the proportion of tartaric and oxalic acids and the strength of the lime-juice are considerably reduced, and bisulphate of potash, oil of vitriol, and cream of tartar are introduced instead.

The ordinary operations practised on calico in this style of work to obtain a white figure are the following:

1. The cloth is padded or printed with the solution of the mordant for the ground, and is immediately dried by being drawn either through the hot-flue or over steam boxes;

2. After a moderate ageing, the calico, without being washed, is imprinted by the roller with the discharging paste, which immediately dissolves the subsalt formed during the ageing;

3. The calico is next suspended for a day or two in a cool place, not very dry, and if the mordant is peroxide of iron, it is then passed through water heated to about 130° Fahr. and rendered slightly alkaline by the addition of a small quantity of carbonate of soda;*

* The passing of the cloth through a dilute solution of carbonate of soda is sometimes omitted, particularly when alumina is the mordant, in which case a quantity of chalk is added to the dung-beck to neutralize the free acid in the discharger.

4. The cloth is afterwards washed, dunged, and dyed in the vegetable infusion; after which it is cleared by soaping or branning and wincing in solution of chloride of lime in the usual manner. Wherever the acid paste had been applied, the colouring material does not attach itself, in consequence of the removal of the mordant from those parts.

It will be observed that this kind of discharge work is very similar to the resist style, in which an acid paste is first imprinted on the cloth to prevent the attachment of a mordant subsequently applied to the whole surface of the cloth (see page 395); the only difference between the two styles consisting in the order of applying the acid and the mordant. The best whites are no doubt generally procured by the resist style; as it is easier for an acid to prevent the attachment of a mordant in an insoluble form, than to dissolve it, when once precipitated.

To procure a white design on a black ground, such as the annexed specimen (No. 22), by the discharge

No. 22.

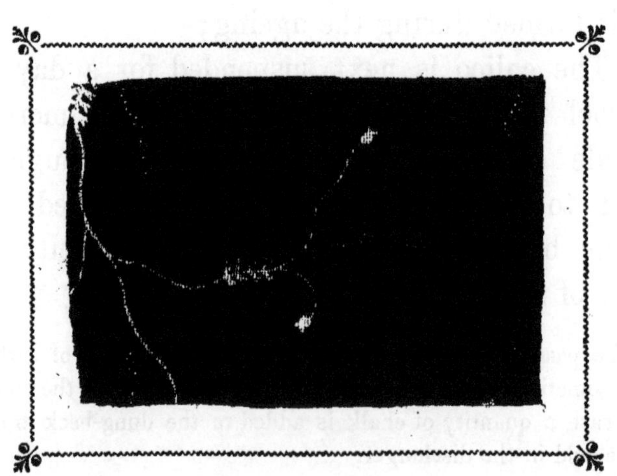

of the mordant, the cloth may be treated in the following manner:

1. Pad or print the calico with a mixture of equal measures of iron liquor of spec. grav. 6° Twad., and red liquor of 8° Twad., thickened with starch or British gum;

2. Dry over the steam boxes, age, and apply a discharger composed of tartaric acid, sulphuric acid, and lime-juice, thickened with British gum;

3. Pass the cloth through warm water mixed with chalk;

4. Dye in decoction of logwood, mixed with a little bran and dung;

5. Wash, clear the white by branning, rinse and dry.

The following method of producing white and blue figures on a purple or chocolate ground presents an example of the combination of such a style as the above with the indigo resist style:

1. The white calico is padded with red liquor;

2. After the cloth has been aged for a short time, the thickened acid discharger is applied by the cylinder to all the parts intended to be blue or white;

3. After hanging for twenty-four hours, the calico is dunged, dyed in the madder-beck, and cleared by branning;

4. On the parts of the white spots which are intended to remain white, the sulphate of zinc resist for the indigo-vat, such as the mixture described at page 403, is imprinted;

5. After the cloth is dried, it is dipped into the

blue-vat and exposed to the air; then washed at the dash-wheel, and dried.

The white figure is here produced through the discharge of the aluminous mordant by the acid, and by the action of the sulphate of zinc resist on the indigo: the blue figure is produced by the indigo on the white spots to which the resist was not applied, and the purple or chocolate ground results from the mixture of the indigo with the madder red.*

A discharger for one mordant is sometimes mixed with the solution of another mordant on which it exerts no action, so that the mordant in the discharger becomes attached to the cloth, on the spots from which the previous mordant is removed. Thus, subacetate of iron may be separated from a piece of calico, and alumina imparted in its place, by applying to the mordanted cloth a mixture of red liquor with protochloride of tin. In this manner, a red figure on a violet or lilac ground is sometimes produced, the cloth being first covered with weak iron liquor, then dried, printed with the mixture of red liquor and protochloride of tin, dunged,

* A simpler and better method of obtaining the same effect is by the "chocolate ground neutral style," the principle of which is described at page 410. The cloth is first printed with the white cupreous resist (mixed with a free acid, when a very well defined figure is required), and afterwards with the chocolate resist (page 411) for the ground, the parts required in blue being left white. The cloth is then aged, drawn once through the blue-vat, washed, dunged, dyed with madder, and cleared by branning. This interesting style of work is very little practised at present, it being superseded by the cheaper but much less permanent steam blue and steam sapan chocolate, of which a specimen is introduced at page 374.

dyed in the madder-beck, and cleared in the usual manner. To obtain a white figure as well as the red, the mordanted cloth should be also printed with lemon-juice, or with a mixture of lemon-juice and sulphuric acid.

The use of a mixture of protochloride of tin and red liquor as a red *resist* for iron liquor, with the view of producing the same effect, has before been adverted to (pp. 397, 398). One of these two processes is almost always followed whenever a figure in madder red is required on a ground of madder purple or black.

In a few ingenious processes related to this kind of work, for producing coloured figures on coloured grounds, an acid solution of protochloride of tin is applied as a kind of discharger to a cloth dyed uniformly with peachwood, quercitron or madder, by means of an iron mordant.

By mutual decomposition, the protochloride of tin in the discharger and the peroxide of iron on the cloth give rise to chloride of iron and peroxide of tin, or rather, the oxide of tin intermediate between the protoxide and peroxide. The chloride of iron, being soluble, is removed by washing, but the insoluble oxide of tin remains attached to the fibre, and combines with the colouring principle previously united to the oxide of iron. This double decomposition of oxide of iron and protochloride of tin may be made subservient to the production of a red figure on a black ground. For this purpose the cloth is first covered with iron liquor, then dyed to a black in decoction of peachwood, and afterwards printed with the acid solution of the protochloride. Wherever the salt of tin is applied, the colour of the cloth changes from black to red, through the transference of the colouring principle of the peachwood from oxide of iron to oxide of tin. In a similar manner, a red figure may be obtained on a ground of madder purple or lilac, and by substituting quercitron for madder or peachwood, a yellow figure is produced on a drab ground. The iron liquor employed in these processes should not be stronger than 3° or 4° Twad.

6. CHINA BLUE STYLE.

The style of calico-printing by which the China blue prints are produced is an interesting modification of the topical style. These prints are distinguished by having blue figures, usually of two or three different depths of colour, associated with white, as the specimen No. 23.

To produce such a pattern, the bleached calico is subjected to the following operations. It is first printed, either by the block or cylinder, with a mixture of indigo, orpiment (sulphuret of arsenic), sulphate of iron or iron liquor, gum or starch, and water; the proportions of gum or starch and water being varied according to the depth of colour required. After being printed, the calico is suspended in a dry atmosphere for a day or two, and stretched in perpendicular folds on a rectangular wooden frame, suspended by pulleys and a rope from the ceiling of the apartment. The frame with the cloth is then dipped

No. 23.

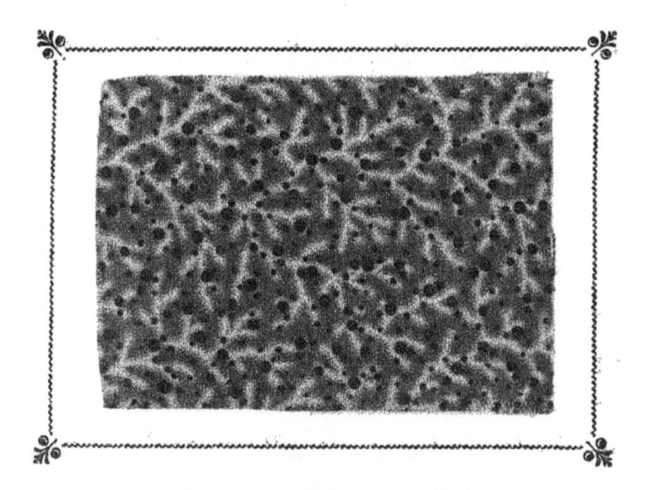

in a certain order into the three following liquids: No. 1, milk of lime; No. 2; solution of copperas; No. 3, solution of caustic soda. These liquids are contained in three adjacent stone cisterns, the tops of which are on a level with the ground: the usual dimensions of the cisterns are, eight or nine feet in length, four feet in depth, and three feet in width.

Into the vats No. 1 and No. 2, the calico is dipped several times alternately, with exposure to the air for a short time between each dip; it is not dipped so frequently into the vat No. 3, and the dipping in this always immediately follows No. 2. By these operations, the insoluble indigo-blue applied to the surface of the cloth becomes converted into indigotin which is dissolved and transferred to the interior of the fibres, where it is precipitated in the original insoluble form.

The chemical changes which take place in these successive operations are rather complicated, but admit, nevertheless, of a satisfactory explanation. By the successive immersions in milk of lime and solution of sulphate of iron, protoxide of iron comes to be precipitated on the surface of the cloth. This protoxide of iron, with the assistance of the lime, reacts on the indigo imprinted on the calico, converting it, through the intervention of water, into indigotin, which dissolves in the lime-water, and the solution is absorbed into the pores of the cloth. On exposure to the air, the indigotin absorbs oxygen, and insoluble indigo-blue is deposited within the fibre in a fixed state. The protoxide of iron produced in the subsequent alternate immersions into the sulphate of iron vat and into the lime-vat, acts only on the indigo which still remains on the surface of the calico, not having free access to the indigo within the fibre; and as the alternate dippings are continued (up to a certain number), so the proportion of indigo on the surface diminishes, and that of the indigo within the fibre increases.

The orpiment contained in the pigment printed on the cloth seems to act chiefly by increasing the density of the mixture, thus preventing its ready disintegration and removal by the various liquids to which the cloth is exposed. It also assists, probably, in deoxidizing the indigo-blue in conjunction with the lime.

By these operations the whole surface of the cloth becomes impregnated with a considerable quantity of oxide of iron, to remove which, the cloth is plunged (being still on the frame) into a fourth similar cistern containing sulphuric acid of about the spec. grav. 5° Twad. (1·025.) It is afterwards washed in clean water, and then again brightened in dilute sulphuric acid. Lastly, the clearing of the white ground is sometimes completed by exposing the cloth to warm soap-water.

The following method of preparing the China blue mixture of different shades is described by M. Thillaye, in his useful work on calico-printing.* The materials employed are,

15¾ pounds of indigo, in coarse powder,
3¾ pounds of orpiment,
22 pounds of copperas, and
9½ gallons of water, or water and gum-water.

The indigo, orpiment, copperas, and four gallons and a half of the water are well ground together in a mill for three days; the mass is then removed, and the mill is washed with a gallon of water which is added to the mixture. The remaining four gallons of water are afterwards added; but if a very thick blue is required, as much strong gum-water is introduced instead. From this mixture, which may be called No. 1, several lighter shades are procured by diluting it with water or gum-water in the following order:

No.	*Quantity by measure of No. 1.*		*Quantity by measure of water or gum-water.*
1	1	mixed with	0
2	11	,,	1
3	10		2
4	8	,,	4
5	6		6
6	4		8
7	2	..	10
8	2		12
9	2		14
10	2		16
11	2	,,	18
12	2	,,	20

* *Manuel du fabricant d'Indiennes*, Paris, 1834.

To produce a small single blue figure, the mixture No. 5, thickened with starch, may be applied by the block, and No. 4, thickened with gum, by the roller.

For two different blues, applied by the block, there may be used, 1°, the mixture No. 4, thickened with starch; and 2°, No. 9, thickened with gum.

For three different blues, applied by the block, there may be taken, 1°, the mixture No. 5, thickened with starch; 2°, No. 7, thickened with starch; and 3°, No. 10, thickened with gum.

The mixture described by M. Thillaye is not exactly the same as that commonly employed in this country. Instead of copperas, the Lancashire printers generally use iron liquor, and British gum instead of common gum; they also take little more than half as much orpiment as is directed in the receipt of M. Thillaye. The following proportions of the materials will probably be found to form a convenient mixture:

16 pounds of ground indigo,
5 or 6 gallons of strong iron liquor,
2 pounds of orpiment, and
British gum and water sufficient to make 8 gallons.

When required for use, this mixture, which contains two pounds of indigo to the gallon, may be diluted with water or gum-water in the order following:

No.	Quantity by measure of above mixture.	Quantity by measure of water or gum-water.	Quantity of indigo in one gallon of the mixture. lbs.	oz.
1	1	0	2	0
2	1	$\frac{1}{2}$	1	$5\frac{1}{4}$
3	1	$\frac{3}{4}$	1	$3\frac{1}{4}$
4	1	1	1	0
5	1	2	0	$10\frac{3}{4}$
6	1	3	0	8
7	1	5	0	$5\frac{1}{4}$
8	1	7	0	4
9	1	9	0	$3\frac{1}{5}$
10	1	12	0	$2\frac{1}{2}$
11	1	16	0	$1\frac{1}{4}$

The darkest of the two shades of blue in the specimen of this style of printing (page 431) was produced from a mixture containing one pound of indigo to the gallon (as No. 4), and the lighter from a mixture containing three ounces of indigo to the gallon (No. 9). Both were thickened with two pounds of British gum per gallon, and were applied at once by the two-colour machine.

The milk of lime for dipping China blue prints may be prepared by mixing two hundred pounds of lime with a thousand gallons of water. When in constant use, the lime-vat requires to be replenished twice daily, both with lime and water.

The strength of the solution of copperas is varied from $3\frac{1}{2}°$ Twad. (1·017) to 6° Twad. (1·030), it being regulated more by the quantity of the figure in the pattern than by the depth of colour required. The kind of copperas generally preferred for this purpose is that technically known as "green copperas," which contains a small quantity of free sulphuric acid. The superiority of this variety of copperas merely consists in the comparative slowness with which it becomes peroxidized or "rusty" by exposure to the air. The copperas-vat does not require replenishing quite so frequently as the lime-vat, and the cistern need not be emptied for six months or longer. The bottom and sides of the cistern become lined with a dense crystalline deposit of oxide of iron and sulphate of lime, as hard as the cistern itself.

The strength of the solution of caustic soda may vary from 6° to 9° Twaddell (1·030 to 1·045). It is made in the usual manner by carbonate of soda and quick-lime.

The order of dipping the frame into the three cisterns is as follows:

1. Dip in the first vat (lime) for ten minutes.
   Drain for five minutes.
2. Dip in the second vat (copperas) for ten minutes.
   Drain for five minutes.
3. Dip in the first vat for ten minutes.
   Drain for five minutes.
4. Dip in the second vat for ten minutes.
   Drain for five minutes.
5. Dip in the third vat (soda) for ten minutes.
   Drain for five minutes.
6. Dip in the second vat for ten minutes.
   Drain for five minutes.
7. Dip in the first vat for ten minutes.
   Drain for five minutes.
8. Dip in the second vat for ten minutes.
   Drain for five minutes.

9. Dip in the first vat for ten minutes.
   Drain for five minutes.
10. Dip in the second vat for ten minutes.
   Drain for five minutes.
11. Dip in the third vat for ten minutes.
   Drain for five minutes.

The addition of a small quantity of nitrate of lead to the China blue mixture, when iron liquor and not copperas is used, is said to impart considerable vivacity to the colour; but I am not aware of its being usual to make this addition, unless with a view of producing a green instead of a blue design, when the cloth, after having been dipped as above and cleared in dilute sulphuric acid, is winced in a solution of bichromate of potash in order to produce chrome-yellow. This is by no means an advantageous process for obtaining a green figure in the China blue style, as the lime and soda vats are apt to become so highly charged with oxide of lead as to deposit that oxide on the white parts, which consequently become yellow when the cloth is exposed to the bichromate. A better method is to add red liquor and perchloride of tin to the China blue mixture, and to dip in the three vats in the usual manner; after which the cloth is cleared in very dilute sulphuric acid, dyed in decoction of quercitron mixed with size, cleared by branning, and lastly, winced in a dilute solution of alum to brighten the green.

Printing of mousselin de laines, silks, chalis, &c.—The fixation of colouring matters on fabrics of silk and wool is commonly effected by the process of steaming. No mineral colouring material, with the exception of Prussian blue, is applied to these tissues; nor is it usual to impart to them colouring matters from infusions of vegetable or animal productions by the madder style, except in a few difficult processes of silk printing. These fabrics were formerly printed entirely by the block, but latterly the roller and the press-machine (page 341) have been substituted.

The colour mixtures for de laines, which are formed of cotton and wool, should be of such a nature as to afford an uniform deposit of colouring matter on both the animal and vegetable fibre. These mixtures are sometimes composed of two distinct bases, one capable of attaching itself firmly to the wool, the other to the cotton. Thus, a preparation sometimes used for imparting a blue colour to de laines, is a mixture of the steam blue for cotton (see page 374) with indigo-paste or soluble blue (sulph-indigotate of potash) for the wool. In a particular kind of fancy dyeing, the woollen thread only is dyed, and the cotton is afterwards perfectly bleached by exposing the dyed de laines to a dilute solution of bleaching powder. The cotton thread is sometimes dyed of another colour, either before or after the dyeing of the woollen thread. Before being printed, de laines are always impregnated with peroxide of tin, from two different solutions applied consecutively as already described (see page 261). The steaming of the printed de laines, which is performed either by the column or chamber, usually lasts about three-quarters of an hour; but the time varies according to the quantity of acid in the mixtures, and to the manner in which the steam is applied. With a considerable quantity of acid, the fibres become weakened if the process is prolonged, and a shorter time is required with the column than with the chamber.

In general, the only difference between the composition of the mixtures for steam colours for woollen goods and those for cotton goods (page 368) is that the former contain more free acid than the latter, or that the colouring matter is held in solution more strongly in the former than in the latter. Whether the mordant is perchloride of tin, protochloride of tin, or alum, a considerable quantity of tartaric or oxalic acid is almost always introduced. The most vivid colours are generally obtained by protochloride of tin, with either oxalic or tartaric acid.

The brilliant steam blue distinguished when on woollen goods as "royal blue," is formed through the decomposition of hydroferrocyanic acid, as before explained. The composition of the mixture printed on the cloth is much the same as the steam blue for cotton (page 374), but is more concentrated, and perchloride of tin is introduced instead of alum. The solution of yellow prussiate of potash, which should contain not less than three pounds of the prussiate in a gallon, is mixed with sufficient tartaric acid to precipitate the whole of its potash as bitartrate of potash (cream of tartar), which may be separated and employed in the preparation of tartaric acid.

The best mixture for a steam scarlet for woollen goods, is made of cochineal liquor, gall liquor, protochloride of tin, and oxalic acid. It may be thickened with gum, if for blotches or grounds; and with starch, if for small figures. The best mordant for producing different shades of yellow with berry liquor is a mixture of alum and red liquor; and with decoction of quercitron bark, a mixture of red liquor and oxalic acid, or else alum alone.

The only preliminary operations to which silken cloth intended to be printed, is commonly subjected, are, 1°, boiling in a solution of soap and carbonate of soda to remove the "gum;" 2°, passing through dilute sulphuric acid; and 3°, washing and drying. Some printers recommend the steeping of the silk in a solution of alum, after it is taken out of sulphuric acid, but this is by no means a common practice. The processes for printing and dyeing silks according to the madder style, are very similar to those for cottons: the thickened mordant is first applied; the piece is then dried, aged for a couple of days, winced in bran-water, dyed in the hot vegetable decoction, and cleared by being winced in boiling bran-water. In some styles of work, the silk is afterwards soaped, impregnated with a solution of tin, and, lastly, passed through very dilute sulphuric acid.

The madder style of printing and dyeing is rather difficult of execution on silken cloth, and is consequently not much practised on this variety of textile fabrics. The common method of ornamenting silk with different coloured designs is by means of steam colours, a great variety of which may be imparted of sufficient fixity to bear all ordinary exposure to deteriorating influences. With the exception of the preliminary operations previous to the application of the colour-mixture, the treatment of silks in this style of work is the same as that of cottons.

A remarkable style of printing and dyeing is largely practised on silken and woollen goods, which possesses not the smallest resemblance to any one of the varied processes to which cotton goods are subjected. This is the *mandarining style*, by which a yellow or orange colour may be communicated to the silk or wool by exposing the stuff to the action of nitric acid. The colour proceeds from a peculiar substance formed through the decomposition by the acid of a portion of the fibre of the cloth itself. On the parts intended to be preserved white, the contact of the acid with the cloth is prevented by the application of a resist paste composed of resin and suet.

Taking, in the first place, the simplest illustration of this style of work, that is, the production of a white design on an orange ground, the operations practised on the cloth are the following.

The silk, having been first cleaned from its resinous coating in the usual manner, is printed on the parts which are to remain white with the fatty resist of resin and suet. A quantity of this mixture is kept in a melted state near at hand, and a portion is occasionally laid over a piece of woollen cloth stretched on a frame which forms the top of a copper chest. Steam is admitted into the chest by a pipe in order to preserve the resist in a liquid state. When the silk is ready to receive the resist, the printer heats his block, takes up some of the resist from the frame, and applies it immediately to the silk by a light blow with a mallet. The block is instantly removed, to prevent it from adhering to the silk.

When the printing of the whole piece is completed, the silk is passed through dilute nitric acid, obtained by adding from one to two parts of water to one part of aquafortis of commerce; the acid is heated as high as possible without endangering the solidity of the resist paste, the melting of which would evidently be attended with serious inconveniences. The silk should not remain in the acid longer than one minute.

The nitric acid is contained in a sand-stone or an earthenware trough, which is placed within a copper or wooden box to serve as a water-bath: the heat applied is that of steam. A reel is placed on each side of the trough to guide the silk as it enters and leaves the acid.

On being withdrawn from the acid, the silk is immediately washed in a stream of cold water; after which, the resist is cleared away, and the orange brightened by wincing the silk in boiling soap-water to which a little soda is added. The piece is lastly washed in cold and hot water successively, and dried.

Such is the method of obtaining a white figure on an orange ground. Coloured figures are obtained in this style of work by a variety of ingenious and elegant processes, applicable, like that just described, to chalis as well as silks. One or two examples will suffice as illustrations.

White and blue figures on an orange ground may be procured in the following manner. The piece is first printed with the resinous resist, to preserve the parts which are to remain white from contact with the acid. It is next dyed blue in the indigo-vat in the ordinary manner, washed and dried, and the resist is then printed on those parts which are to remain blue. The next process is the mandarining or passing through the nitric acid; by which the indigo is destroyed on all parts of the cloth except where the resist is applied, the cloth thus acquiring a ground of orange. The piece is lastly washed, and the orange brightened by a boil in soap-water with a little alkali.

An orange ground having been applied to the cloth by such a process as the preceding, the colour of the ground may be afterwards modified or completely altered by dipping the cloth in some dyeing liquid, the figures being still protected, if necessary, by the resinous resist. For example, a design in white and blue figures on a green ground may be imparted to a chali by simply dipping the piece, after having been treated as above for blue and white figures on an orange ground, in the blue-vat, previons to the removal of the resinous resist from the blue and white figures. White, blue, and orange figures on a green ground may be obtained by imprinting the resinous resist on the silk or chali after the mandarining process, and before the second immersion in the blue-vat.

An orange figure on a blue ground is sometimes produced by printing the nitric acid thickened with British gum upon the cloth previously dyed in the blue-vat. The piece is afterwards exposed to the action of steam, and is lastly boiled in soap-water to brighten the orange.

Superficial as the preceding account of calico-printing processes may and must appear to those who are acquainted, both practically and theoretically, with all the details of this beautiful art, it will probably be found sufficiently minute and exact to substantiate the claim of calico-printing to be considered not only one of the most important, but the most ingenious and refined of all the chemical arts. From the great variety of processes and of materials employed, almost every principle in theoretical chemistry receives an application or illustration in some one or other of the operations of the calico-printer. It has thus happened that several interesting discoveries in theoretical chemistry, made in the experimental laboratory, have actually been anticipated by the printer, from observations made in the print-works and dye-house.

INDEX.

D.

T.

LONDON:
Printed by S. & J. Bentley, Wilson, and Fley,
Bangor House, Shoe Lane.

Lightning Source UK Ltd.
Milton Keynes UK
UKHW012133061118
331891UK00009B/426/P